SpringerBriefs in Electrical and Computer Engineering

Control, Automation and Robotics

Series editors

Tamer Başar
Antonio Bicchi
Miroslav Krstic

W0192898

More information about this series at http://www.springer.com/series/10198

Luciano Pandolfi

Distributed Systems with Persistent Memory

Control and Moment Problems

 Springer

Luciano Pandolfi
Dipartimento di Scienze Matematiche
  "G.L. Lagrange"
Politecnico di Torino
Torino, Italy

ISSN 2192-6786   ISSN 2192-6794   (electronic)
ISBN 978-3-319-12246-5   ISBN 978-3-319-12247-2   (eBook)
DOI 10.1007/978-3-319-12247-2

Library of Congress Control Number: 2014951738

Mathematics Subject Classification (2010): 45K05, 93B03, 93B05, 93C22

Springer Cham Heidelberg New York Dordrecht London

Printed on acid-free paper

Springer is part of Springer Science+Business Media (www.springer.com)

Preface

This booklet is devoted to the study of controllability of systems with persistent memory, which are important in engineering applications, for example, to visco-elasticity, non-Fickian diffusion and thermal processes with memory. Earlier studies of controllability are due to Leugering, see [63–65], and by now controllability under the action of boundary deformation has been studied with several different methods while controllability under boundary traction has been less considered. We give an overview of some of the ideas used in this kind of study for systems of "hyperbolic" type, i.e. finite velocity of signal propagation (the "parabolic" case is far less studied, see [8, 37, 41–43]): operator methods (introduced by Belleni-Morante in [10] and used for control problems in [74]) are in Chap. 2; a moment method approach to controllability is in Chap. 5; a circle of ideas introduced by Kim in [52] and which relays on the observation inequality is in Chap. 6.

To familiarize the readers with the subject, Chap. 1 treats a very simple example; Chap. 4 recalls known properties of the controllability of the (memoryless) wave equation and Chap. 3 presents the results we shall need on Riesz sequences, stressing the applications to the solution of the moment problems encountered in "hyperbolic" type systems.

We note that "controllability" as studied here is not state controllability as defined in [50] since the "state" of a system with persistent memory takes into account the entire past evolution of the system, see [18, 19, 25], and so state controllability is clearly impossible: we study the possibility of hitting a target at a given time. Section 5.4 shows an application of this notion of controllability to a source identification problem.

Finally, we mention problems that we are not going to study. First, we note that new possibilities arise when studying systems with persistent memory: the flux of heat and the temperature or, analogously, the stress and the deformation or the velocity of deformation, are only weakly related and it makes sense to understand how strong the relation is. This problem can be recast in the form of a control problem and leads to the study of the controllability of the pairs (temperature/flux), (deformation/stress) and (velocity of deformation/stress), which is a novelty of memory systems, see [4, 3, 80]. We shall always assume that the memory kernels are smooth. Approximate controllability for kernels with (mild) singularities has been studied for example in [31, 40, 49]. Finally we cite [29, 97], where controllability is studied using Carleman estimates.

This book is addressed to people interested in control theory, with basic knowledge of functional analysis. In order to state clearly the results that are needed, the final section of Chap. 1 recalls general notions while more special properties are explicitly recalled when needed.

The solutions to the problems proposed in every chapter can be downloaded from the author's WEB page at the address http://calvino.polito.it/~lucipan/ricerca.html.

The author is a member of the *Gruppo Nazionale per l'Analisi Matematica, la Probabilità e le loro Applicazioni (GNAMPA)* of the *Istituto Nazionale di Alta Matematica* (INdAM) and of the project *Groupement de Recherche en Contrôle des EDP entre la France et l'Italie (CONEDP)*.

Contents

Chapter 1
An Example

1.1 The Goal of This Chapter and Preliminaries

In this preliminary chapter, in order to get a feeling of the properties of distributed systems with persistent memory and to introduce the key definitions, we study the very simple example of an unbounded viscoelastic string (laying on a half line) controlled at its end. We shall see in Chap. 2 that this example is simple but significant. Relying on known properties of the purely elastic string, we introduce the definition of the solutions and we study a control problem.

In order to help the reader and to fix the notations, the final section of this chapter recalls the key definitions and notions that we shall use. Readers with a feeling of the subject can skip this chapter since this example is not used in the rest of the book (it is used in a few problems).

We recall a few properties of the string equation on a half line (see [92])

$$u'' = u_{xx} + F(x, t), \quad t > 0, \quad x > 0 \tag{1.1}$$

with initial and boundary conditions

$$u(x, 0) = u_0(x), \quad u'(x, 0) = v_0(x), \quad u(0, t) = f(t). \tag{1.2}$$

The solution is the sum of three functions:

$$u(x, t) = u_1(x, t) + u_2(x, t) + u_3(x, t) \tag{1.3}$$

where:

- $u_1(x, t)$ is the contribution of the initial conditions,

$$u_1(x, t) = \frac{1}{2} [\phi(x + t) + \phi(x - t)] + \frac{1}{2} \int_{x-t}^{x+t} \psi(s) ds \tag{1.4}$$

© The Author(s) 2014
L. Pandolfi, *Distributed Systems with Persistent Memory*, SpringerBriefs in Control, Automation and Robotics, DOI 10.1007/978-3-319-12247-2_1

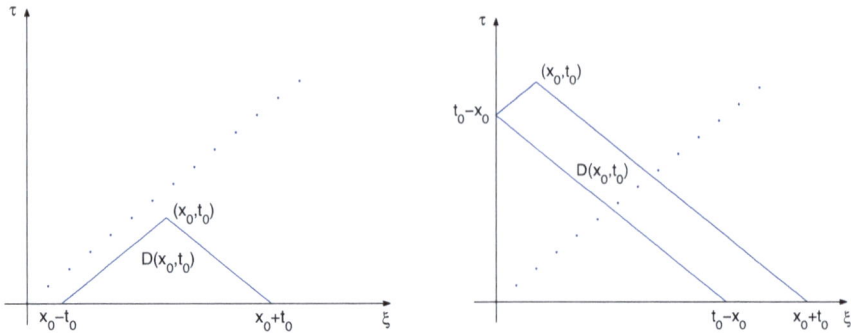

Fig. 1.1 The domains $D(x_0, t_0)$ when $x_0 > t_0$ (*left*) and $x_0 < t_0$ (*right*)

where $\phi(x)$ and $\psi(x)$ are the odd extensions to $\mathbb{R}$ of $u_0(x)$ and $v_0(x)$.
- $u_2(t)$ is the contribution of the boundary term f,

$$u_2(x, t) = f(t - x)\mathbf{H}(t - x) \tag{1.5}$$

where $\mathbf{H}(t)$ is the Heaviside function, $\mathbf{H}(t) = 1$ if $t \geq 0$ and $\mathbf{H}(t) = 0$ if $t < 0$. This is an abuse of notation, since f is defined only for $t > 0$. The notation wants to indicate that we consider the *extension* of $f(t)$, which is zero for $t < 0$.
- $u_3(t)$ is the contribution of the affine term F,

$$u_3(x, t) = \frac{1}{2} \int_{D(x,t)} F(\xi, \tau)d\xi\, d\tau = \frac{1}{2} \int_0^t \int_{|x-t+\tau|}^{x+t-\tau} F(\xi, \tau)d\xi\, d\tau. \tag{1.6}$$

In the plane (ξ, τ), the region $D(x_0, t_0)$ is identified by the inequalities

$$0 \leq \tau \leq t_0, \quad |(x_0 - t_0) + \tau| \leq \xi \leq x_0 + t_0 - \tau.$$

More explicitly, we have two cases: if $x_0 \geq t_0 \geq 0$ then $D(x_0, t_0)$ is the triangle of the plane (ξ, τ) identified by the characteristics

$$\tau - t_0 = \xi - x_0, \quad \tau - t_0 = -(\xi - x_0) \tag{1.7}$$

(Fig. 1.1, left). Otherwise, if $0 \leq x_0 \leq t_0$, then $D(x_0, t_0)$ is the quadrilateral in Fig. 1.1, right.

We are interested in the following consequences of (1.4)–(1.6):
- let $u_0(x) = 0$, $v_0(x) = 0$ for $x > X > 0$. Then $u_1(x, t) = 0$ if $x > X + t$: signals propagate with *finite speed* (equal to 1).
- In order to have a differentiable function $u(x, t)$, which can be substituted into the equation, we must require certain regularity of the data, for example $u_0 \in C^2$,

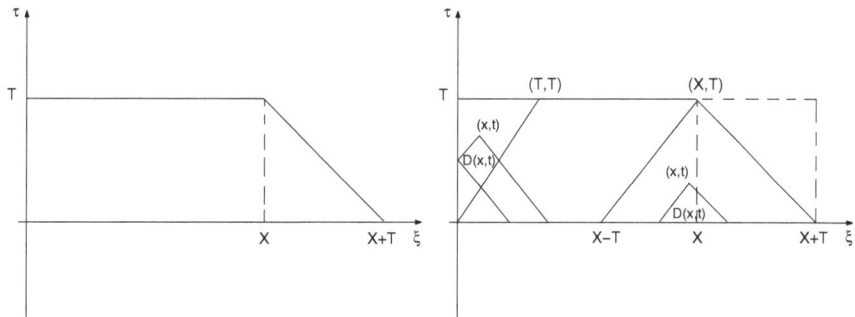

Fig. 1.2 The trapezium and its extension to the rectangle

$v_0 \in C^1$, $f \in C^2$ and $F \in C^1$, and also in this case we do not get a regular solution u unless we have certain compatibility conditions among the data.

- the function u, sum of (1.4)–(1.6), makes sense also if such regularity conditions do not hold. In particular, we have the following properties (see Sect. 1.3 for the definition of the space H_0^1). We fix any $T > 0$ and any $X > T$. We consider the restriction to $0 \le t \le T$ and $0 \le x \le X + T$ of the function $u(x, t)$ in (1.3). We consider the trapezium in Fig. 1.2, left, then extended to the rectangle $(0, X + T) \times (0, T)$ (right). We have:

  – Let $(u_0, v_0) \in H_0^1(0, X) \times L^2(0, X)$, then extended with zero for $x > X$. The transformation $(u_0, v_0) \mapsto u_1 \in C([0, T]; L^2(0, X + T))$ is continuous (even more, $u_1(x, t) \in H_0^1(0, X + T)$ for every $t \in [0, T]$).
  – Let $F(x, t) = 0$ on the right of the line joining $(X - T, 0)$ and (X, T) (Fig. 1.2, right). The transformations $L^1(0, T; L^2(0, X + T)) \ni F \mapsto u_3 \in C([0, T]; L^2(0, X + T))$ is continuous (even more, $u_3(x, t) \in H_0^1(0, X + T)$ for every $t \in [0, T]$).
  – the transformation $f \mapsto u_2$ is continuous from $L^2(0, T)$ to $C([0, T]; L^2(0, X + T))$ (but in general u_2 *does not* take values in $H_0^1(0, X + T)$).
  – $u(x, t) \in C([0, T]; L^2(0, X + T))$ depends continuously on the data u_0, v_0, f and F taken in the spaces specified above.

In conclusion, when the boundary input f is merely of class L^2, the case we shall be interested in, we are forced to read u in $C([0, T]; L^2(0, X))$. Continuous dependence on the data, as specified above, shows that the function u in (1.3) is the limit in $C([0, T]; L^2(0, X + T))$ of "regular" solution. For this reason it is still considered to be a "solution" of Eq. (1.1) even if it is not differentiable. It is called a MILD SOLUTION.

A final comment on the FINITE PROPAGATION SPEED: let $F = 0$, $u_0 = 0$ and $v_0 = 0$ so that

$$u(x, t) = u_2(x, t) = f(t - x)\mathbf{H}(t - x).$$

Let $f(t) \neq 0$ for $t \in [a, b]$ and $f(t) = 0$ for $t \in [0, a)$ (when $a > 0$) and for $t > b$. Then,

$$u(x, t) = 0 \quad \text{when } t < x + a:$$

the signal generated at time a covers the distance x_0 in time $x_0 + a$: *it propagates with velocity* 1. We have also $u(x_0, t) = 0$ when $t > x_0 + b$; i.e., the signal acts on the position x_0 during the interval of time $[x_0 + a, x_0 + b]$. In a picturesque way, we can say *that the signal leaves no memory of itself after it passed over the point x_0.*

1.2 The System with Persistent Memory

We study the following equation:

$$w'' = w_{xx} + \int_0^t K(t - s)w(x, s)ds, \quad x > 0, \ t > 0. \tag{1.8}$$

The function $K(t)$ is continuous. For simplicity we assume null initial conditions. So, we impose

$$w(x, 0) = 0, \quad w'(x, 0) = 0, \quad w(0, t) = f(t). \tag{1.9}$$

Equation (1.1) is the *string equation associated with the equation with memory* (1.8) and Eq. (1.8) is the same as (1.1), the role of $F(x, t)$ being taken by

$$\int_0^t K(t - s)w(x, s)ds.$$

So, we have the following representation formula for the solutions:

$$w(x, t) = f(t - x)\mathbf{H}(t - x) + \frac{1}{2} \int_{D(x,t)} \left[\int_0^\tau K(\tau - s)w(\xi, s)ds \, d\xi \right] d\tau. \tag{1.10}$$

In order to obtain this formula, we considered the equation with memory as a perturbation of the memoryless string equation. We note:

- For every time t, the integral on the right hand side of (1.10) depends only on $w(\xi, s)$ with $s \leq t$, as it is to be expected from (1.8).
- the properties of the string equation suggest $w \in C([0, T]; L^2(0, X))$, for every $T > 0$ and $X > 0$.

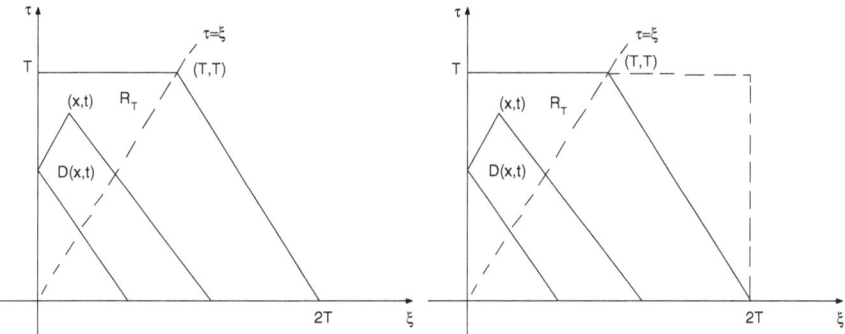

Fig. 1.3 The trapezium R_T (*left*) and the corresponding rectangle $(0, 2T) \times (0, T)$ (*right*)

Now we prove that Eq. (1.10) is solvable: we prove that Eq. (1.10) admits a solution $w(x, t) \in L^2_{loc}(0, +\infty; L^2_{loc}(0, +\infty))$, which "depends continuously" on f (in the sense specified below). Time continuity is then seen directly, from (1.10).

We fix any $T > 0$ and we note that

$$t \le T, \quad x \le 2T - t \implies D(x, t) \subseteq R_T$$

where R_T is the trapezium in Fig. 1.3 (left). We study $w(x, t)$ in R_T for every fixed $T > 0$ (note that we are not asserting that w is defined *only* in R_T).

We fix a parameter $\gamma > 0$ (to be specified later on) and we rewrite Eq. (1.10) in the equivalent form

$$y(x, t) = \psi(x, t) + \frac{1}{2} \int_{D(x,t)} e^{-\gamma(t-\tau)} \int_0^\tau \left[e^{-\gamma(\tau-s)} K(\tau - s) \right] y(\xi, s) ds \, d\xi \, d\tau \tag{1.11}$$

where

$$y(x, t) = e^{-\gamma t} w(x, t), \quad \psi(x, t) = e^{-\gamma t} f(t - x) \mathbf{H}(t - x). \tag{1.12}$$

A function w solves (1.10) if and only if $y = e^{-\gamma t} w$ solves (1.11). So, it is enough that we study Eq. (1.11): we must study the restriction of $y(x, t)$ to R_T.

We prove that the transformation

$$y \mapsto \frac{1}{2} \int_{D(x,t)} e^{-\gamma(t-\tau)} \int_0^\tau \left[e^{-\gamma(\tau-s)} K(\tau - s) \right] y(\xi, s) ds \, d\xi \, d\tau \tag{1.13}$$

is linear and continuous on $L^2(R_T)$ to itself and that it is a contraction, i.e., its norm is *strictly less then* 1, when γ is sufficiently large. This implies that Eq. (1.11) admits a solution in $L^2(R_T)$, and the solution is unique, see Sect. 1.3.

In order to compute the norm in $L^2(R_T)$, we extend $y(x, t)$ with 0 to the rectangle $\tilde{R}_T = (0, 2T) \times (0, T)$ (see Fig. 1.3, right). Let $\tilde{y}(x, t)$ be this extension and let $|K(t)| < M$ for $t \in [0, T]$. Then, we have:

$$
\left| \int_{D(x,t)} e^{-\gamma(t-\tau)} \int_0^\tau \left[e^{-\gamma(\tau-s)} K(\tau - s) \right] y(\xi, s) ds\, d\xi\, d\tau \right|^2
$$

$$
\leq \left| \int_{\tilde{R}_T} e^{-\gamma(t-\tau)} \int_0^\tau \left[e^{-\gamma(\tau-s)} |K(\tau - s)| \right] |\tilde{y}(\xi, s)| ds\, d\xi\, d\tau \right|^2
$$

$$
\leq \left(\int_{\tilde{R}_T} e^{-2\gamma t} d\xi\, d\tau \right) \left(\int_{\tilde{R}_T} \left[\int_0^\tau e^{-2\gamma s} K^2(s) ds \right] \left[\int_0^\tau \tilde{y}^2(\xi, s) ds \right] d\xi\, d\tau \right)
$$

$$
\leq \frac{M^2 T^2}{2\gamma^2} \left(\int_{\tilde{R}_T} \tilde{y}^2(\xi, s) d\xi\, ds \right) = \frac{M^2 T^2}{2\gamma^2} |y|^2_{L^2(R_T)}.
$$

So, the norm of the transformation is strictly less then 1 if γ is sufficiently large, as we wanted and we can state:

Theorem 1.1 *Let $T > 0$ and let $f(t) \in L^2(0, T)$. Then, Eq. (1.11) admits a unique solution $y(x, t) \in L^2(R_T)$. The transformation $f \mapsto y$ from $L^2(0, T)$ to $L^2(R_T)$ is continuous.*

Remark 1.1 It is easily seen that the integral term in (1.11) is a continuous function of (x, t). So, *the memory term does not add discontinuities to the solution* and if $f = 0$ for $t > T$ then $y \in C([0, T]; L^2(0, X))$ for every $X > 0$.

1.2.1 Finite Propagation Speed

It is important to understand that also in the presence of memory, signals propagate with FINITE SPEED. We consider system (1.11)–(1.12) so that only the boundary signal f enters the system. We prove that the input f does not affect a point in position x if $t < x$:

$$
t < x \implies y(x, t) = 0.
$$

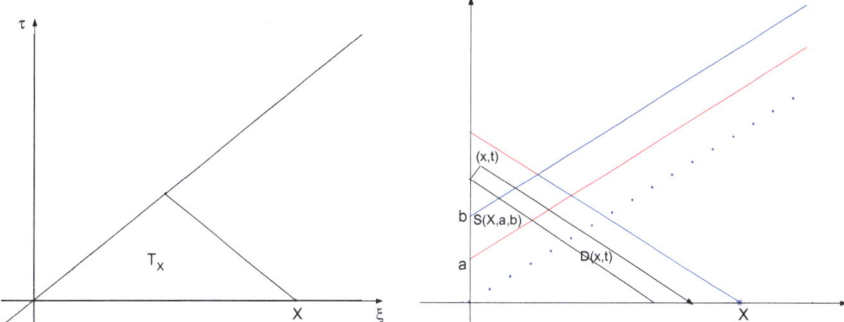

Fig. 1.4 The domain T_X (*left*) and the domains $Q(X, b)$, $Q(X, a)$ and $S(X, a, b)$ (*right*)

We fix any $X > 0$ and we consider the domain (a triangle)

$$T_X = \{(\xi, \tau) \; : \; 0 < \tau < \xi < X - \tau\},$$

which is represented in Fig. 1.4, left. Note that $D(x, t) \subseteq T_X$ for every $(x, t) \in T_X$. Hence, the integral operator in (1.11) is also a transformation in $L^2(T_X)$ (and it is a contraction for γ large). We observe that $\psi = 0$ on T_X. The integral operator (1.11) being a contraction in $L^2(T_X)$, the equation has a unique solution. The fact that $\psi = 0$ on T_X implies that $y = 0$. This proves that *signals propagates with finite velocity* also in the presence of memory.

Now we consider the effect of a forcing term f which is zero for $0 \leq t \leq a$ and for $t > b$.

Again we fix any $X > 0$ and this time we consider the two parallels to the line $\tau = \xi$, from $(0, a)$ and from $(0, b)$, and their intersection with the line $\xi = X - \tau$ (assume $b < X$). We get two quadrilaterals, respectively, $Q(X, a)$ and $Q(X, b)$, represented in Fig. 1.4, right. The quadrilateral $Q(X, a)$ is the one below the line $\tau = \xi + a$ and $Q(X, b)$ is below the line $\tau = \xi + b$.

Let $S(X, a, b) = Q(X, b) \setminus Q(X, a)$, i.e., the strip between the lines $\tau = \xi + a$ and $\tau = \xi + b$.

It is clear that if $(x, t) \in Q(X, a)$, then $D(x, t) \subseteq Q(X, a)$. The same argument as above show that $y(x, t) = 0$ in $Q(X, a)$. Instead, if $(x, t) \in Q(X, b) \setminus Q(X, a)$, i.e., if (x, t) is in the strip, in general it will be $y(x, t) \neq 0$.

In conclusion, if $f(t) = 0$ for $t < a$, then $y(x, t) = 0$ in $Q(X, a)$ for every $X > 0$. This is the same as for the memoryless string equation. But, let us consider now a point (x, t), which is *above* the line $\tau = \xi + b$, as in Fig. 1.4 (right). It is clear that $D(x, t)$ will intersect the strip $S(x, t)$, and then we can represent

$$\int_{D(x,t)} \int_0^\tau \mathrm{d}s \, \mathrm{d}\xi \, \mathrm{d}\tau = \int_{D(x,t) \cap S(X,a,b)} \int_0^\tau \mathrm{d}s \, \mathrm{d}\xi \, \mathrm{d}\tau + \int_{D(x,t) \setminus S(X,a,b)} \int_0^\tau \mathrm{d}s \, \mathrm{d}\xi \, \mathrm{d}\tau.$$

In general, these integrals will not be zero: the effect of the persistent memory is to introduce a "persistent affine term" on the right hand side of (1.11) and we cannot expect y to become zero after the effect of the boundary term f has been "forgotten" by the solution ψ of the memoryless string equation: *once the signal has reached a point of abscissa x, we have to expect that the memory of the signal will persist forever in the future.* It is indeed so, as shown in the important example examined in Sect. 2.6.3.

1.2.2 A Formula for the Solutions and a Control Problem

We give a representation for the solutions of (1.11) (i.e., of (1.8)–(1.9)) and we use it to solve a control problem. We recall finite propagation speed: $y(x, t) = 0$ when $x > t$.
   We fix a time T and we confine ourselves to study the solutions in the square

$$Q = \{(x, t), \quad 0 < t < T, \quad 0 < x < T\}.$$

Since we know that $y(x, t) = 0$ when $x > t$, the integral operator in (1.13) can be considered as a linear and continuous transformation in $L^2(Q)$ (in the computations of the integrals we can extend y to the right of Q with $y = 0$). We already proved that the norm of the transformation can be taken as small as we wish, by a proper choice of γ. *We fix γ so to have the norm less then* $1/4$ (and note that the integral has a factor $1/2$ in front). We elaborate the integral operator. Let

$$\tilde{K}(t, \tau, s) = e^{-\gamma(t-s)} K(\tau - s).$$

Then, we have

$$\int_{D(x,t)} \int_0^\tau \tilde{K}(t, \tau, s) y(\xi, s) ds \, d\xi \, d\tau$$

$$= \int_0^t \int_0^{x+t-\tau} \mathbf{H}(\xi - |x - t + \tau|) \int_0^\tau \tilde{K}(t, \tau, s) y(\xi, s) ds \, d\xi \, d\tau$$

$$= \int_0^t \int_0^{x+t} \left\{ \mathbf{H}(x + t - \tau - \xi) \mathbf{H}(\xi - |x - t + \tau|) \right.$$

$$\times \left[\int_0^t \mathbf{H}(\tau - s) \tilde{K}(t, \tau, s) y(\xi, s) ds \right] \Big\} d\xi \, d\tau$$

$$= \int_0^t \int_0^{x+t} G_1(t, x, s, \xi) y(\xi, s) d\xi \, ds \tag{1.14}$$

where

$$G_1(t, x, s, \xi) = \int_0^t \mathbf{H}(x + t - \tau - \xi)\mathbf{H}(\xi - |x - t + \tau|)\mathbf{H}(\tau - s)\tilde{K}(t, \tau, s)d\tau.$$

Now, we use finite propagation speed so that $y(\xi, s) = 0$ for $\xi > s$, in particular for $\xi > t$, and we get

$$y(x, t) = \psi(x, t) + \frac{1}{2}\int_0^t\int_0^t G_1(t, x, s, \xi)y(\xi, s)d\xi\,ds = \psi + \frac{1}{2}Gy$$

where G is the integral transformation in $L^2(Q)$

$$Gy = \int_0^t\int_0^t G_1(t, x, s, \xi)y(\xi, s)d\xi\,ds.$$

Then, we have (see Sect. 1.3)

$$y = \psi + \sum_{k=1}^{+\infty}\frac{1}{2^k}G^k\psi = \psi + \frac{1}{2}G\left(\sum_{k=0}^{+\infty}\frac{1}{2^k}G^k\psi\right)$$

(the series converges since $\|G\| < 1/4$ thanks to the choice of γ). This is the representation we wanted to achieve. In particular, we see that

$$y(x, T) = \psi(x, T) + \frac{1}{2}\int_0^T\int_0^T G_1(T, x, s, \xi)\left(\sum_{k=0}^{+\infty}\frac{1}{2^k}G^k\psi\right)d\xi\,ds. \qquad (1.15)$$

The norm of the following transformation in $L^2(0, T)$ is less then $8/7$:

$$\psi \mapsto \tilde{G}\psi = \left(\sum_{k=0}^{+\infty}\frac{1}{2^k}G^k\psi\right).$$

We interpret $\psi(x, t) = e^{-\gamma t}f(t - x)\mathbf{H}(t - x)$ (see (1.12)) as a control, to be used in order to force y to reach (or to hit) a prescribed target at time T: we fix any $\xi(x) \in L^2(0, T)$ and we require that the following equality hold for $x \in (0, T)$:

$$\xi(x) = y(x, T) = \psi(x, T) + \frac{1}{2}\int_0^T\int_0^T G(T, x, s, \xi)\left(\tilde{G}\psi\right)d\xi\,ds. \qquad (1.16)$$

This equation in the unknown ψ is uniquely solvable, since the transformation in $L^2(0, T)$

$$\psi \mapsto \frac{1}{2} \int_0^T \int_0^T G(T, x, s, \xi) \left(\tilde{G} \psi \right) ds \, d\xi$$

is a contraction: its norm is less then $1/7$.

So, *the control problem we posed is (uniquely) solvable.* Of course, the control is unique on the interval $(0, T)$. Its extension in the future is arbitrary.

Remark 1.2 The simple model we have studied has its interest, see Remark 2.3.

We studied controllability of the deformation, but not joint controllability of the pair (deformation and velocity of deformation). Controllability of the pair is not possible in an unbounded string, not even in the memoryless case. In fact

$$u(x, T) = f(T - x) \implies u'(x, T) = f'(T - x):$$

deformation and velocity of deformation are not independent. Compatibility conditions for controllability of the pair (in fact, controllability to zero from certain non zero initial conditions) have been recently given in [26] in the memoryless case.

We shall see that the couple can be controlled in a finite string, thanks to the reflection of the signals generated at one end from the other end of the string.

Analogously, it is also impossible (in the memoryless case) to control both the deformation and the stress. In fact, the derivation of the string equation depends on Hooke law: the stress is *proportional* to $u_x(x, T)$, which is determined by $u(x, t)$. So, controllability of the pair $(u(x, T), u_x(x, T))$ for the (memoryless) string equation is impossible even in a finite string. Memory instead gives more degrees of freedom, as proved in [4, 3, 80].

1.3 Background of Functional Analysis

We assume basic knowledge of functional analysis, as can be found in standard books. We refer to [1, 7, 11, 17, 51, 72]. Here, we recall general properties which will be currently used. More specific properties will be recalled at appropriate places.

A linear operator L acting among Banach spaces is continuous if and only if it is BOUNDED, i.e., if and only if its norm is finite: $\|Lx\| < +\infty$ where

$$\|Lx\| = \sup_{|x|=1} |Lx|.$$

The Banach space of the continuous linear operators from H to K (both Banach spaces) is denoted $\mathscr{L}(H, K)$ ($\mathscr{L}(H)$ when $H = K$). When K is the scalar field, in practice either $\mathbb{R}$ or $\mathbb{C}$, it is denoted H' (the DUAL SPACE of H). If $\chi \in H'$ then the value taken by χ on the element $h \in H$ (the PAIRING of the space and its dual) is denoted[1]

$$\langle\!\langle h, \chi \rangle\!\rangle.$$

Most of the time we shall work with a Hilbert space H, whose norms and inner products are denoted $|\cdot|$ and $\langle \cdot, \cdot \rangle$ (an index identifies the space, e.g., $|\cdot|_H$, when needed for clarity).

It is known that the dual space of a Hilbert space H can be represented by H itself, and we do this identification for the spaces L^2 and l^2, but not for the Sobolev spaces introduced below.

1.3.1 Weak Topology in a Hilbert Spaces

NORM CONVERGENCE (also called STRONG CONVERGENCE) of a sequence $\{x_n\}$ to x_0 in a Hilbert space H, denoted $x_n \to x_0$, means that $|x_n - x_0|_H \to 0$. This is equivalent to the following property: for every $\varepsilon > 0$ there exists N_ε such that

$$n > N_\varepsilon \text{ and } |h| \le 1 \implies |\langle x_n, h \rangle - \langle x_0, h \rangle| < \varepsilon.$$

We say that $\{x_n\}$ CONVERGES WEAKLY to x_0 in a Hilbert space H when for every $h \in H$ we have

$$\lim \langle x_n, h \rangle = \langle x_0, h \rangle$$

i.e., for every $\varepsilon > 0$ we can find $N_{\varepsilon,h}$ such that

$$n > N_{\varepsilon,h} \implies |\langle x_n, h \rangle - \langle x_0, h \rangle| < \varepsilon.$$

We stress the dependence of $N_{\varepsilon,h}$ on h even if $|h| \le 1$ in the case of weak convergence.
To denote weak convergence we use one of the following notations:

$$\text{w} - \lim x_n = x_0, \quad \text{or} \quad x_n \rightharpoonup x_0.$$

The properties we need of the weak convergence are:

- any weakly convergent sequence is bounded.
- any bounded sequence in a Hilbert space has weakly convergent subsequences.
- if $T \in \mathcal{L}(H, K)$ and if $x_n \rightharpoonup x_0$ in H then $T x_n \rightharpoonup T x_0$ in K.

---

[1] We shall introduce the space $\mathscr{D}(\Omega)$, which is not a Banach space. Also in this case the value on $\phi \in \mathscr{D}(\Omega)$ of an element χ of the dual $\mathscr{D}'(\Omega)$ is denoted $\langle\!\langle \phi, \chi \rangle\!\rangle$.

- A property of the norm is as follows: if $x_n \rightharpoonup x_0$ then, as we noted, $\{|x_n|\}$ is a bounded sequence of real numbers, and so it has convergent subsequences. *Every* convergent subsequence $\{|x_{n_k}|\}$ has the following property:

$$|x_0| \leq \lim |x_{n_k}|.$$

This property is shortly written

$$|x_0| \leq \liminf |x_n| \tag{1.17}$$

and it is called the WEAK LOWER SEMICONTINUITY of the norm.

1.3.2 Operators and Resolvents

Let A be a linear operator from a Hilbert space H to a Hilbert space K. The graph of A is

$$\mathscr{G}(A) = \{(x, Ax), \ x \in \mathrm{dom}\,A\}.$$

The set $\mathscr{G}(A)$ is a linear subspace of $H \times K$. If it is a *closed subspace*, then the operator A is a CLOSED OPERATOR. If an operator is closed and invertible, its inverse is closed too. Important properties are:

- if the operator A is closed and everywhere defined on H then it is continuous.
- if the operator A is closed, surjective, and invertible, then its inverse A^{-1} is continuous.

If it happens that the operator $A: H \mapsto K$ has dense domain in H, then it is possible to define its (Hilbert space) ADJOINT OPERATOR A^*, acting from K to H, as follows: A^*k is that element $z \in H$ for which the following equalities hold:

$$\langle z, h \rangle_H = \langle k, Ah \rangle_K \quad \forall h \in \mathrm{dom}\,A.$$

By definition, $\mathrm{dom}\,A^*$ is the sets of those $k \in K$ for which there exists an element $z \in H$ which satisfies these equations (density of the domain of A implies that z is unique). So, A and A^* satisfy

$$\langle A^*k, h \rangle_H = \langle k, Ah \rangle_K \quad \forall h \in \mathrm{dom}\,A, \ \forall k \in \mathrm{dom}\,A^*.$$

The operator A is SELFADJOINT when it acts from H to itself and $A = A^*$ (note that this implies $\mathrm{dom}\,A = \mathrm{dom}\,A^*$).

We shall need the following properties:

- any adjoint operator is closed. So, in particular, a selfadjoint operator is closed.
- if A is continuous then A^* is continuous (and conversely).
- if the operator A is closed with dense domain then A^* has dense domain.

The next result is proved for example in [11, Chap. II.7] and in [35, Sects. II.3 and II.4]:

Theorem 1.2 *Let A be a linear closed operator with dense domain in a Hilbert space H and image in a Hilbert space K. Then, we have:*

1. *the operator A is invertible with bounded inverse if and only if there exists $m > 0$ such that $m|x|_H \le |Ax|_K$ for every $x \in \mathrm{dom}\,A$.*
2. *If A^* is surjective then A is invertible and A^{-1} is bounded and conversely.*
3. *the operator A is surjective if and only if $(A^*)^{-1}$ is continuous i.e. if and only if there exists $m_1 > 0$ such that $m_1|k|_K \le |A^*k|_H$.*
4. *A is continuous and boundedly invertible if and only if its adjoint is continuous and boundedly invertible.*

The RESOLVENT SET $\rho(A)$ of the operator A acting from H to itself is the set of those numbers $\lambda \in \mathbb{C}$ such that (I denotes the identity operator, $I x = x$):

- $(\lambda I - A)x = 0$ if and only if $x = 0$, and so the RESOLVENT OPERATOR $(\lambda I - A)^{-1}$ exists.
- the image of $(\lambda I - A)$ is dense in H and $(\lambda I - A)^{-1}$ is continuous.

If $(\lambda I - A)x = 0$ has a *non zero* solution x then λ is called an EIGENVALUE of A and x is an EIGENVECTOR.

We conclude with the following definitions:

- Let $t \mapsto R(t)$ be a map defined from (a, b) to $\mathscr{L}(H, K)$. The map is STRONGLY CONTINUOUS when $t \mapsto R(t)h$ is a K-valued continuous function for every $h \in H$.
- let $A \colon H \mapsto H$ be boundedly invertible and let $u(t)$ be an H-valued function, which takes values in $\mathrm{dom}\,A$. We say that $u \in C^k(0, T; \mathrm{dom}\,A)$ when $Au(t) \in C^k(0, T; H)$.

1.3.3 Compact Operators

We noted that $T \in \mathscr{L}(H, K)$ transforms weakly convergent sequences to weakly convergent sequence. A K-valued operator T defined on H (both Hilbert spaces) is a COMPACT OPERATOR when it transforms *every* weakly convergent sequence in H to a norm convergent sequence in K. Of course, replacing $\{x_n\}$ with $\{(x_n - x_0)\}$ (when $x_n \rightharpoonup x_0$), in order to prove that T is compact we can prove

$$y_n \rightharpoonup 0 \text{ implies } T y_n \to 0.$$

It is easily seen that that *every compact operator is continuous* and that *the composition of a continuous operator with a compact operator is compact.*

If it happens that $H = K$, a Hilbert space, then we can study the eigenvalues of the compact operator T, i.e. we can study the equations

$$Tx = \lambda x, \quad x \neq 0, \quad \lambda \in \mathbb{C}.$$

We have the following properties:

- there exist only a finite number of eigenvectors, which correspond to a *non zero* eigenvalue.
- every $\lambda \neq 0$ which is not an eigenvalue belongs to the resolvent set.
- if λ is an eigenvalue of the compact operator T, then for every eigenvector x_0 corresponding to λ we can consider the chain of equations

$$Tx_n = \lambda x_n + x_{n-1}, \quad n > 1.$$

The sequence $\{x_n\}$ (whose first element is the eigenvector x_0) is a JORDAN CHAIN of T and it turns out that every Jordan chain, which corresponds to an eigenvalue $\lambda \neq 0$ of a compact operator has *finite length* i.e., finitely many non zero elements.
- Let A be densely defined on H, linear and with values in H. If it is selfadjoint and if its resolvent $(\lambda I - A)^{-1}$ is a compact operator[2], then the operator A has a sequence of normalized (i.e., of norm equal 1) eigenvectors, which is an orthonormal basis of H (and every Jordan chain of A has length 1).

1.3.4 Fixed Point and Equations

We confine ourselves to consider linear equations of the form

$$Tx = f \quad \text{i.e., in equivalent form putting } T = I + L, \quad x + Lx = f. \quad (1.18)$$

Here, I is the identity operator on a Hilbert space H and T and L are linear and continuous operators from H to H.

We give two conditions, which ensure unique solvability of this equation, and also that the solution x depends continuously on f.

BANACH FIXED POINT THEOREM states that Eq. (1.18) admits a unique solution for every f if L is a CONTRACTION, i.e. if (strict inequality is crucial here!)

$$\|T - I\| = \|L\| = \sup_{|x|=1} |Lx| < 1.$$

---

[2] for one value of λ. But this implies that $(\lambda I - A)^{-1}$ is compact for every $\lambda \in \rho(A)$.

The unique solution is a linear and continuous function of f and it is given by the VON NEUMANN FORMULA:

$$x = \sum_{n=0}^{+\infty} (-1)^n L^n f.$$

A consequence of this is:

Theorem 1.3 *For every bounded and boundedly invertible operator $\mathcal{T}_0$, there exists a number $\varepsilon = \varepsilon(\mathcal{T}_0) > 0$ such that $\mathcal{T}_0 + \mathcal{T}$ is invertible when $\|\mathcal{T}\| < \varepsilon$.*

The second condition is as follows:

Theorem 1.4 *The equation $(I + L)x = f$ admits a unique solution x for every f when L is a compact operator and $I + L$ is invertible, i.e., -1 is not an eigenvalue of L. Under this condition, $x = (I + L)^{-1} f$ depends continuously on f.*

1.3.5 Volterra Integral Equations

The convolution $f * g$ of f and g is defined as follows:

$$h(t) = (f * g)(t) = \int_0^t f(t - s)g(s)\mathrm{d}s, \quad t \in (0, T).$$

A Volterra integral equation (of convolution type) on a Hilbert space H is

$$w(t) + \int_0^t R(t - s)w(s)\mathrm{d}s = f(t). \tag{1.19}$$

The function $R(t)$ is the KERNEL of the Volterra integral equation. We assume that it takes values in $\mathcal{L}(H)$ for every $t > 0$ and that $t \mapsto R(t)$ is strongly continuous. Let furthermore $R(t)R(\tau) = R(\tau)R(t)$ for every t and τ in $[0, T]$ and $f(t)$ belong to $L^2(0, T; H)$.

The notation R^{*k} denotes iterated convolution of R with itself:

$$R^{*1} = R, \quad R^{*(k+1)} = R * R^{*k}.$$

Equation (1.19) admits a unique solution $w(t)$ in $L^2(0, T; H)$, which depends continuously on $f \in L^2(0, T; H)$ (moreover, if $f(t)$ is continuous then also $w(t)$ is continuous). The solution w is given by the PICARD FORMULA

$$w = f - \sum_{k=1}^{+\infty}(-1)^{k-1}R^{*k} * f \qquad (1.20)$$

(the series converges in $L^2(0, T; H)$). Let

$$L = \sum_{k=1}^{+\infty}(-1)^{k-1}R^{*k}.$$

Then, we can write

$$w = f - L * f.$$

The function $L(t)$ is called the RESOLVENT KERNEL of $R(t)$ and it is identified by the fact that it is the unique solution of the Volterra integral equation

$$L + R * L = R.$$

See [16, 85] for the theory of Volterra integral equations in Hilbert spaces.

Let us consider the following Volterra integro-differential equation (where now N, f and w are scalar valued, $N(t)$ is continuous and $f(t)$ square integrable):

$$w' = 2\alpha w + \int_0^t N(t - s)w(s)ds + f(t), \quad w(0) = w_0.$$

We are interested in a formula for the solution $w(t)$, which exists, is unique and depends continuously on t since integrating both the sides we get a Volterra integral equation. Let $z(t)$ solve

$$z'(t) = 2\alpha z(t) + \int_0^t N(t - s)z(s)ds, \quad z(0) = 1.$$

Then, $w(t)$ is given by the following VARIATION OF CONSTANTS formula:

$$w(t) = z(t)w_0 + \int_0^t z(t - s)f(s)ds. \qquad (1.21)$$

The verification is as follows:

$$w(t) - z(t)w_0 = z(0)w(t) - z(t)w(0) = \int_0^t \frac{d}{ds} z(t-s)w(s)ds$$

$$= \int_0^t \left\{ \left[-2\alpha z(t-s) - \int_0^{t-s} N(t-s-r)z(r)dr \right] w(s) \right.$$

$$\left. + z(t-s) \left[2\alpha w(s) + \int_0^s N(s-r)w(r)dr + f(s) \right] \right\} ds$$

$$= \int_0^t z(t-s)f(s)ds.$$

1.3.6 Test Functions and Distributions

Let N and n_i be nonnegative integers. We use the notations

$$\alpha = (n_1, \ldots, n_N), \text{ and } D^\alpha = \frac{\partial^{n_1 + \cdots n_N}}{\partial x_1^{n_1} \partial x_2^{n_2} \ldots \partial x_N^{n_N}}, \quad |\alpha| = n_1 + \cdots + n_N.$$

The number $|\alpha|$ is the "length" of the string α.

The special case of the first partial derivative respect to x_r will be denoted D_r,

$$D_r \phi = \frac{\partial \phi}{\partial x_r}.$$

Let Ω be an open set in any (finite) number N of dimensions. The TEST FUNCTIONS on Ω are the functions $\phi \in C^\infty(\Omega)$ with compact support in Ω.

The linear space of the test functions is denoted $\mathscr{D}(\Omega)$ and a sequence ϕ_n converges to ϕ_0 in $\mathscr{D}(\Omega)$ when both the following conditions hold:

- the supports of every ϕ_n and of ϕ_0 are contained in a compact subset $K \subseteq \Omega$.
- for every α, $D^\alpha \phi_n \to D^\alpha \phi_0$ uniformly on Ω.

The dual space $\mathscr{D}'(\Omega)$ is the space of the linear functionals χ on $\mathscr{D}(\Omega)$, which are continuous in the sense that[3]

$$\phi_n \to \phi_0 \implies \langle\!\langle \phi_n, \chi \rangle\!\rangle \to \langle\!\langle \phi_0, \chi \rangle\!\rangle.$$

---

[3] as already told, the pairing of $\mathscr{D}'(\Omega)$ and $\mathscr{D}(\Omega)$ is denoted $\langle\!\langle \phi, \chi \rangle\!\rangle$: $\chi(\phi) = \langle\!\langle \phi, \chi \rangle\!\rangle$.

By definition,

$$\chi_n \to \chi \text{ in } \mathscr{D}'(\Omega)$$

when (compare with the definition of weak convergence)

$$\lim \langle\!\langle \phi, \chi_n \rangle\!\rangle = \langle\!\langle \phi, \chi \rangle\!\rangle \quad \forall \phi \in \mathscr{D}(\Omega).$$

The elements of $\mathscr{D}'(\Omega)$ are called DISTRIBUTIONS.

The operators D^α can be defined also on $\mathscr{D}'(\Omega)$, as follows:

$$\langle\!\langle \phi, D^\alpha \chi \rangle\!\rangle = (-1)^{|\alpha|} \langle\!\langle D^\alpha \phi, \chi \rangle\!\rangle.$$

It is clear that D^α from $\mathscr{D}(\Omega)$ to itself and also from $\mathscr{D}'(\Omega)$ to itself is continuous. We call REGULAR DISTRIBUTIONS those distributions χ_f defined by

$$\langle\!\langle \phi, \chi_f \rangle\!\rangle = \int_\Omega f(x)\phi(x)\mathrm{d}x \tag{1.22}$$

where $f \in L^1_{\text{loc}}(\Omega)$ (unicity of the representation is easily seen). Distributions which are not regular exist, for example the DIRAC DELTA $\phi \mapsto \langle\!\langle \phi, \delta \rangle\!\rangle = \phi(0)$.

For every $f \in L^1_{\text{loc}}(\Omega)$ we can compute $D^\alpha \chi_f$. In general, $D^\alpha \chi_f$ is not a regular distribution (for example, $D\mathbf{H} = \delta$). If it is a regular distribution, $D^\alpha \chi_f = \chi_g$, then the function g is called the DERIVATIVE IN THE SENSE OF THE DISTRIBUTIONS of the function f and it is shortly denoted $D^\alpha f$.

Finally, we say that $\chi = 0$ on a subregion Ω' of Ω when $\langle\!\langle \phi, \chi \rangle\!\rangle = 0$ for every $\phi \in \mathscr{D}(\Omega)$ whose support is contained in Ω'. The complement of the largest region over which $\chi = 0$ is the SUPPORT of χ.

It is a fact that if $K \subseteq \Omega$ and K is compact, then there exist (infinitely many) functions $f \in \mathscr{D}(\Omega)$ such that $f(x) = 1$ on K. Hence, if $\phi \in C^\infty(\Omega)$ (possibly not zero on the boundary), then $f(x)\phi(x) \in \mathscr{D}(\Omega)$ so that we can compute (with $\chi \in \mathscr{D}'(\Omega)$)

$$\langle\!\langle f\phi, \chi \rangle\!\rangle. \tag{1.23}$$

It is a fact:

Lemma 1.1 *If χ has compact support $K \subsetneq \Omega$ and $\phi \in C^\infty(\Omega)$, then the number (1.23) does not depend on the function $f \in \mathscr{D}(\Omega)$, which is equal 1 on K and it is simply denoted $\langle\!\langle \phi, \chi \rangle\!\rangle$.*

In this sense, distributions with compact support can be applied to C^∞ functions, and not only to the functions of $\mathscr{D}(\Omega)$ and furthermore we have:

if $\phi \in C^\infty(\Omega)$, χ with compact support then $\langle\!\langle D_i\phi, \chi \rangle\!\rangle = -\langle\!\langle \phi, D_i\chi \rangle\!\rangle.$ (1.24)

Fig. 1.5 The definition of
"region with C^2 boundary"

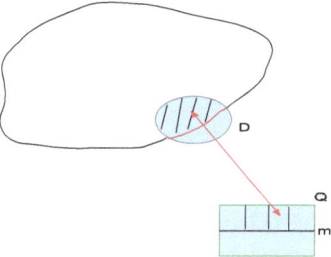

1.3.7 Sobolev Spaces

We shall work in a *bounded* region $\Omega \subseteq \mathbf{R}^N$ whose *boundary is of class* C^2 and which lays on one side of its boundary. The meaning is as follows: every point of $\partial\Omega$ has a neighborhood D which is transformed to the interior of a "cube" Q by a transformation F, which has the following properties, illustrated in Fig. 1.5. In this figure m is a plane which split the "cube" in two parallelepipeds Q_+ and Q_-.

- The function F is of class C^2, and its jacobian J is not zero.
- the image of open set $D \cap \Omega$ is contained Q_+, say the part above the "median" m.
- the set $D \cap \partial\Omega$ is transformed to the median m.

Using $J \neq 0$ we can represent $\partial\Omega \cap D$ as $\xi = \psi(\eta)$ (if D is "small").

- (ξ, η) belongs to $D \cap \Omega$ if and only if $\xi > \psi(\eta)$.

This last property is crucial for the definition of the trace.

We define $H^m(\Omega)$ as the spaces of those $L^2(\Omega)$ functions whose derivatives in the sense of distributions up to the order m included are regular distributions, identified by square integrable functions. The spaces $H^m(\Omega)$ are Hilbert spaces with the norm

$$|\phi|^2_{H^m(\Omega)} = \sum_{|\alpha| \leq m} \int_\Omega |D^\alpha \phi(x)|^2 \mathrm{d}x. \tag{1.25}$$

The space $C^\infty(\overline{\Omega})$ is dense both in $H^1(\Omega)$ and in $H^2(\Omega)$ (smoothness of $\partial\Omega$ is used here).

Furthermore, we define $H_0^m(\Omega)$ as the closure of $\mathscr{D}(\Omega)$ in $H^m(\Omega)$ (in the case we are interested in, that $\Omega \subseteq \mathbf{R}^N$ is a bounded region, $H_0^m(\Omega) \neq H^m(\Omega)$). The dual space of $H_0^m(\Omega)$ is denoted $H^{-m}(\Omega)$.

On $H_0^1(\Omega)$ the following two norms are equivalent[4]:

---

[4] Two norms on the same linear space are equivalent when a sequence which converges with one of these norms converges, and has the same limit, also with the other one.

$$\left[\int_\Omega |\phi(x)|^2 dx + \sum_{k=1}^N \int_\Omega |D_k\phi(x)|^2 dx\right]^{1/2}, \qquad \left[\sum_{k=1}^N \int_\Omega |D_k\phi(x)|^2 dx\right]^{1/2}.$$

We shall be interested mostly in the spaces $H_0^1(\Omega)$, $H^2(\Omega)$ and $H^{-1}(\Omega)$.

By its very definition, $\mathscr{D}(\Omega)$ is dense in $H_0^1(\Omega)$ and it is clear that when $\phi_n \to \phi$ in $\mathscr{D}(\Omega)$ then we have also $\phi_n \to \phi$ in $H_0^1(\Omega)$. So, any element of $H^{-1}(\Omega)$ is continuous on $\mathscr{D}(\Omega)$: $H^{-1}(\Omega) \subseteq \mathscr{D}'(\Omega)$.

The norm in $H^{-1}(\Omega)$ is

$$|\chi|_{H^{-1}(\Omega)} = \sup\{\langle\!\langle \phi, \chi \rangle\!\rangle, \quad |\phi|_{H_0^1(\Omega)} = 1\} \tag{1.26}$$

so that for every $\chi \in H^{-1}(\Omega)$ and $\phi \in H_0^1(\Omega)$ we have

$$|\langle\!\langle \phi, \chi \rangle\!\rangle| \le |\chi|_{H^{-1}(\Omega)} |\phi|_{H_0^1(\Omega)}. \tag{1.27}$$

Let $\{\chi_{f_n}\}$ be a sequence of regular distributions in $H^{-1}(\Omega)$, such that $\chi_{f_n} \to \chi_0$ in $H^{-1}(\Omega)$. Then, we have

$$\langle\!\langle \phi, \chi_0 \rangle\!\rangle = \lim \langle\!\langle \phi, \chi_{f_n} \rangle\!\rangle = \lim \int_\Omega f_n(x)\phi(x)dx \quad \forall \phi \in H_0^1(\Omega). \tag{1.28}$$

Shortly, we shall write

$$f_n \to \chi_0 \text{ in } H^{-1}(\Omega) \text{ instead of } \chi_{f_n} \to \chi_0. \tag{1.29}$$

Of course, if $f \in \mathscr{D}(\Omega)$ then $\chi_f \in H^{-1}(\Omega)$. We shall say shortly that $\mathscr{D}(\Omega) \subseteq H^{-1}(\Omega)$ and it turns out that $\mathscr{D}(\Omega)$ is dense in $H^{-1}(\Omega)$. Hence, for every $\chi_0 \in H^{-1}(\Omega)$ there exists a sequence of regular distributions χ_{f_n} such that $f_n \in \mathscr{D}(\Omega)$ and for which (1.28) holds.

Let us consider the special case $\Omega = (0, T)$, an interval. Our definition is that the elements of $H^1(0, T)$ are equivalent classes of *square integrable functions*. It is a fact that every equivalent class of $H^1(0, T)$ contains a (unique) continuous element, which furthermore is a.e. differentiable with *square integrable derivative* and this element can be recovered from its derivative, using Lebesgue integration:

$$\phi \in H^1(0, T) \implies \phi(t) = \phi(a) + \int_a^t \phi'(s)ds \tag{1.30}$$

for every a and t in $[0, T]$. This element will be used to identify its class.

The properties of higher order spaces, $(H^2(0, T), H^3(0, T)\dots)$, are easily deduced.

If instead dim $\Omega > 1$, an element $\phi \in H^1(\Omega)$ needs not be represented by a smooth function, not even a continuous function. In spite of this, we would like to define the TRACE of elements of $H^k(\Omega)$ and of their derivatives on $\partial\Omega$.

If $\phi \in C^\infty(\Omega)$ (with continuous derivatives up to $\partial\Omega$) then $\gamma_0\phi = \phi_{|\partial\Omega}$ while $\gamma_1\phi = \partial\phi/\partial\nu$ is the (exterior) normal derivative on $\partial\Omega$. We have (regions whose boundary is of class C^2 and locally on one side of the boundary is crucial here, see [93, Sect. 13.6]):

Lemma 1.2 *The linear transformation γ_0 with values in $L^2(\partial\Omega)$ admits a linear continuous extension to $H^1(\Omega)$ whose kernel is $H_0^1(\Omega)$; the linear transformation γ_1 with values in $L^2(\partial\Omega)$ admits a linear continuous extension to $H^2(\Omega)$.*

If $u \in H^k(\Omega)$ for every k then the equivalent class u contains a (unique) C^∞ function. If $\partial\Omega$ is of class C^∞, then this function and its derivative admit continuous extensions to $\overline{\Omega}$.

1.3.8 Sobolev Spaces and the Laplace Operator

The operator A on $L^2(\Omega)$ defined by

$$\text{dom}\,A = H_0^1(\Omega) \cap H^2(\Omega), \quad A\phi = \Delta\phi = \phi_{x_1 x_1} + \cdots + \phi_{x_N x_N} \tag{1.31}$$

is the LAPLACIAN with HOMOGENEOUS BOUNDARY CONDITIONS. Its key properties are:

- the operator A is selfadjoint (hence it is closed) and surjective, and $0 \in \rho(A)$. So we have $A^{-1} \in \mathscr{L}(L^2(\Omega))$ ($\partial\Omega$ of class C^2 is used here). Furthermore, the operator A from $H^2(\Omega) \cap H_0^1(\Omega)$ to $L^2(\Omega)$ is bounded and boundedly invertible.
- the operator A has compact resolvent. *Hence, $L^2(\Omega)$ has an orthonormal basis of eigenvectors of the operator A.* We denote $\{\phi_n\}_{n>0}$ such a basis. It is possible to choose the functions $\phi_n(x)$ to be *real valued*. We denote $-\lambda_n^2$ the eigenvalue of ϕ_n, $A\phi_n = -\lambda_n^2\phi_n$. We have $\lambda_n^2 > 0$ for every n and $\lim \lambda_n^2 \to +\infty$.
- whether $\lambda_n = \sqrt{\lambda_n^2}$ is taken with positive or negative sign has no influence. We choose the positive sign, i.e. we put

$$\lambda_n = \sqrt{\lambda_n^2} > 0 \quad \text{for} \quad n > 0.$$

- the eigenvalues are not distinct, but each one of them has finite multiplicity. We intend that the eigenvectors are listed in such a way that $\lambda_n \le \lambda_{n+1}$. Once the eigenvalues have been ordered in this way, we know (see [71, p. 192]) that there exist two constants $m_0 > 0$ and M_0 such that

$$m_0 n^{2/N} \le \lambda_n^2 \le M_0 n^{2/N}, \quad N = \dim\Omega.$$

• in terms of expansions in series of $\{\phi_n\}$, we have

$$
\begin{cases}
L^2(\Omega) \ni y = \displaystyle\sum_{n=1}^{+\infty} \alpha_n \phi_n \iff \{\alpha_n\} \in l^2, \\[2ex]
y = \displaystyle\sum_{n=1}^{+\infty} \alpha_n \phi_n \in \operatorname{dom} A \iff \{\lambda_n^2 \alpha_n\} \in l^2, \\[2ex]
Ay = A\left(\displaystyle\sum_{n=1}^{+\infty} \alpha_n \phi_n\right) = \displaystyle\sum_{n=1}^{+\infty} (-\lambda_n^2 \alpha_n) \phi_n.
\end{cases}
\tag{1.32}
$$

The FRACTIONAL POWERS[5] of $(-A)$ can be defined in the obvious way:

$$
y = \sum_{n=1}^{+\infty} \alpha_n \phi_n \in \operatorname{dom}(-A)^\gamma \iff \{\lambda_n^{2\gamma} \alpha_n\} \in l^2, \quad (-A)^\gamma y = \sum_{n=1}^{+\infty} \lambda_n^{2\gamma} \alpha_n \phi_n.
\tag{1.33}
$$

It is easy to see that $(-A)^\gamma$ is boundedly invertible and that $((-A)^\gamma)^{-1} = ((-A)^{-1})^\gamma$.

Note that $\operatorname{dom}(-A)^\gamma$ is a Hilbert space when endowed with the norm

$$
|y|^2_{\operatorname{dom}(-A)^\gamma} = \sum_{n=1}^{+\infty} \left| \lambda_n^{2\gamma} \alpha_n \right|^2
$$

and that, with $\{c_n\} = \{\lambda_n^{2\gamma} \alpha_n\}$ arbitrary in l^2, the elements of $\operatorname{dom}(-A)^\gamma$ have the representation

$$
y \in \operatorname{dom}(-A)^\gamma \iff y = \sum_{n=1}^{+\infty} c_n \frac{\phi_n(x)}{\lambda_n^{2\gamma}}, \quad |y|^2_{\operatorname{dom}(-A)^\gamma} = \sum_{n=1}^{+\infty} |c_n|^2.
\tag{1.34}
$$

It follows that the sequence $\{\phi_n(x)/\lambda_n^{2\gamma}\}$ is an orthonormal basis of $\operatorname{dom}(-A)^\gamma$.

A known fact is (see [32, 59])

$$
\operatorname{dom}(-A)^{1/2} = H_0^1(\Omega)
$$

so that any element of $H_0^1(\Omega)$ has the representation

$$
y = \sum_{n=1}^{+\infty} \alpha_n \phi_n(x) \text{ with } \{\lambda_n \alpha_n\} = \{c_n\} \in l^2, \text{ i.e. } y = \sum_{n=1}^{+\infty} c_n \frac{\phi_n(x)}{\lambda_n}.
$$

---

[5] in fact, the exponent is any positive real number. See below for negative exponents.

The norm in (1.34) with $\gamma = 1/2$ is a norm in $H_0^1(\Omega)$, equivalent to (1.25).

An important consequence is that we can choose $\{\phi_n/\lambda_n\}$ *as an orthonormal basis of* $H_0^1(\Omega)$.

It is also possible to define fractional powers with negative exponents. Of course such operators are continuous and defined on $L^2(\Omega)$. Negative powers can be extended to *larger* spaces, still with image in $L^2(\Omega)$. In fact, we can introduce the spaces

$$\mathcal{H}_{-\gamma} = \left\{ \chi = \sum_{n=1}^{+\infty} \alpha_n \phi_n(x), \quad \sum_{n=1}^{+\infty} \left(\frac{\alpha_n}{\lambda_n^{2\gamma}} \right)^2 < +\infty \right\}, \quad |\chi|_{\mathcal{H}_\gamma}^2 = \sum_{n=1}^{+\infty} \left(\frac{\alpha_n}{\lambda_n^{2\gamma}} \right)^2.$$

The operator $(-A)^{-\gamma}$ can be extended to $\mathcal{H}_{-\gamma}$,

$$(-A)^{-\gamma} \sum_{n=1}^{+\infty} \alpha_n \phi_n = \sum_{n=1}^{+\infty} \frac{\alpha_n}{\lambda_n^{2\gamma}} \phi_n, \quad (-A)^\gamma \in \mathscr{L}(\mathcal{H}_{-\gamma}, L^2(\Omega)).$$

With $\{\delta_n\} = \{\alpha_n/\lambda_n^{2\gamma}\}$ arbitrary element of l^2 we have

$$\begin{cases} \mathcal{H}_{-\gamma} = \left\{ \chi = \sum_{n=1}^{+\infty} \delta_n \left(\lambda_n^{2\gamma} \phi_n \right), \quad \{\delta_n\} \in l^2 \right\}, \\ |\chi|_{\mathcal{H}_{-\gamma}}^2 = \left| \sum_{n=1}^{+\infty} \delta_n \left(\lambda_n^{2\gamma} \phi_n \right) \right|_{\mathcal{H}_{-\gamma}}^2 = \sum_{n=1}^{+\infty} |\delta_n|^2, \\ (-A)^{-\gamma} \left(\sum_{n=1}^{+\infty} \delta_n \left(\lambda_n^{2\gamma} \phi_n \right) \right) = \sum_{n=1}^{+\infty} \delta_n \phi_n. \end{cases} \tag{1.35}$$

Let now $\chi \in \mathcal{H}_\gamma$ and consider any $y \in \mathrm{dom}(-A)^\gamma$, hence represented as in (1.34). Comparing the expressions of the norms (1.34) and in (1.35) is clear that the transformation

$$y \mapsto \sum_{n=1}^{+\infty} c_n \delta_n$$

is linear and continuous on $\mathrm{dom}(-A)^\gamma$. The converse holds too: any element of $(\mathrm{dom}(-A)^\gamma)'$ is represented by a (unique) element of $\mathcal{H}_{-\gamma}$: $(\mathrm{dom}(-A)^\gamma)' = \mathcal{H}_{-\gamma}$.

We are mostly interested in the special case $\gamma = 1/2$: the elements of the dual $H^{-1}(\Omega)$ of $H_0^1(\Omega)$ have the representation

$$\sum_{n=1}^{+\infty} \delta_n \left(\lambda_n \phi_n(x) \right), \quad \{\delta_n\} \in l^2$$

and $\{\lambda_n \phi_n\}$ is an orthonormal basis of $H^{-1}(\Omega)$ (the norm in (1.35) with $\gamma = 1/2$ is equivalent to the norm (1.26)).

As a final observation, we note that the operator $-A$, originally defined on $H^2(\Omega) \cap H_0^1(\Omega)$ with values in $L^2(\Omega) \subseteq H^{-1}(\Omega)$ can be extended to an $H^{-1}(\Omega)$-valued operator defined on $H_0^1(\Omega)$. The extension is still denoted $-A$ and it is given by

$$(-A)\left(\sum_{n=1}^{+\infty} c_n \frac{\phi_n(x)}{\lambda_n}\right) = \sum_{n=1}^{+\infty} c_n \left(\lambda_n \phi_n(x)\right).$$

Known fact: $-A \in \mathcal{L}\left(H_0^1(\Omega), H^{-1}(\Omega)\right)$ is surjective, invertible and with bounded inverse. Hence, $-A$, equivalently, A is an isomorphism between $H_0^1(\Omega)$ and $H^{-1}(\Omega)$.

1.4 Problems[6] to Chap. 1

1.1. Let $\xi(x)$ and $\eta(x)$ be smooth in $[0, +\infty)$ and let f be smooth in $t \geq 0$. Prove that

$$u(x, t) = f(t - x)\mathbf{H}(t - x) + \frac{1}{2}\left[\tilde{\xi}(x + t) + \tilde{\xi}(x - t)\right] + \frac{1}{2}\left[\tilde{\eta}(x + t) - \tilde{\eta}(x - t)\right]$$

(1.36)

is the mild solution of (1.1) with $w(0, t) = f(t)$ and $F = 0$, $w(x, 0) = \xi(x)$, $w'(x, 0) = \eta_x(x)$ if $\tilde{\xi}$ and $\tilde{\eta}$ are suitable extensions of ξ and η. Explain which extension.

Give conditions under which $u(x, t)$ is continuous in $x \geq 0$ and $t \geq 0$.

1.2. It is known that the derivative is a continuous transformation from $L^2(0, +\infty)$ to $H^{-1}(0, +\infty)$. Use formula (1.36) to prove that the transformations $(\xi, \eta, f) \mapsto u$ and $(\xi, \eta, f) \mapsto u_x$ are continuous transformations from $L^2(0, +\infty) \times L^2(0, +\infty) \times L^2(0, +\infty)$ to, respectively, $C([0, T], L^2(0, T))$ and $C([0, T], H^{-1}(0, T))$, for every $T > 0$.

1.3. Study the controllability of system (1.1) when the boundary control f is more regular, for example when $f \in C([0, +\infty))$. What can be deduced for the controllability properties of system (1.8)?

1.4. Consider Eq. (1.1) with $u_0 = 0$, $v_0 = 0$ and $F = 0$, but impose the condition $u_x(0, t) = f(t)$ (Neumann boundary condition). Describe the reachable set at time T when $f(t)$ is continuous or square integrable.

1.5. Use the expression for $u(x, t)$ found in Problem 1.4, to give a formula analogous to (1.15) when the boundary condition of (1.8) is $w_x(0, t) = f(t)$.

1.6. Study the controllability of system (1.8) with the control $w_x(0, t) = f(t)$.

---

[6] Solutions at the address http://calvino.polito.it/~lucipan/materiale_html/P-S-CH-1.

1.7. Prove that if $\{e_n\}$ is an orthonormal basis of a Hilbert space, then $e_n \rightharpoonup 0$.

1.8. Find a sequence $\{x_n\}$ in a Hilbert space such that $x_n \rightharpoonup x_0$ and such that *for no subsequence* we have $\lim |x_{n_k}| = |x_0|$.

1.9. Use the fact that every weakly convergent sequence is bounded to prove that every compact operator is continuous.

1.10. Let H be a Hilbert space and let $\{e_n\}$ be an orthonormal basis of H. Let T be the operator

$$T\left(\sum_{n=1}^{+\infty} \alpha_n e_n\right) = \sum_{n=1}^{+\infty} \frac{\alpha_n}{n} e_n.$$

Prove that this operator is continuous.

1.11. Let $\{e_n\}$ be an orthonormal basis of a Hilbert space H. Note that if $\{x_n\}$ is a sequence in H such that $x_n \rightharpoonup 0$ then $\lim_n \langle x_n, e_k \rangle = 0$ for every k. Use this observation to prove that the linear operator T in problem 1.10 is compact.

1.12. Prove that the Volterra operator on $L^2(0, T)$:

$$x \mapsto \int_0^t F(t - s) x(s) ds$$

with $F(t) \in C^1([0, T])$ is compact.

1.13. Prove that the following operators $A: L^2(0, 1)$ to itself *are not closed:*

1. $Ax = x'$ with domain $C^1(-1, 1)$.
2. $Ax = x''$ with domain $C^2(-1, 1)$.
3. $Ax = x(0)$ with domain $C(-1, 1)$.

1.14. Prove that the following operators $A: L^2(0, 1)$ to itself *are closed:*

1. $A_1 x = x'$ with domain $H^1(0, 1)$.
2. $A_2 x = x''$ with domain $H^2(0, 1) \cap H_0^1(0, 1)$.

Compute the resolvent and the spectrum of these operators.

1.15. Prove that the second operator in Problem 1.14 has compact resolvent.

Chapter 2
The Model and Preliminaries

2.1 The Goal of This Chapter and the System with Persistent Memory

In this book, we study certain results on the controllability of distributed systems with persistent memory (in the final section of this chapter, we show the derivation of the heat equation with memory and the equation of viscoelasticity). In this chapter, we define and give formulas for the solutions of the system under study and we derive their properties, using an operator approach. We define the notion of controllability and prove the key results relevant to the study of control problems. In particular, we prove that signal propagates with finite velocity, as in the case of the (memoryless) wave equation. A special and important case of the equation with persistent memory is the telegrapher's equation. In Sect. 2.6.3, we use this important example to contrast the properties of the systems with memory and those of the (memoryless) wave and heat equations.

We strive for simplicity of presentation and study the simplest significant case[1]:

$$w''(x,t) = 2cw'(x,t) + c_0^2 \Delta w(x,t) + \int_0^t M(t-s)\Delta w(x,s)ds + F(x,t) \quad (2.1)$$

(here $c_0^2 > 0$ and Δ is the laplacian in the space variable x). The initial conditions are

$$w(x,0) = u_0(x), \quad w'(x,0) = v_0(x). \quad (2.2)$$

We assume $x \subseteq \Omega \in \mathbb{R}^d$ ($d \leq 3$ is the case of physical interest).

System (2.1) can be written in the following equivalent form

---

[1] We introduced the velocity term $2cw'$, which has a role in the application of moment methods. A term $c_1 w$ does not make any difference and we ignore it.

© The Author(s) 2014

L. Pandolfi, *Distributed Systems with Persistent Memory*, SpringerBriefs in Control, Automation and Robotics, DOI 10.1007/978-3-319-12247-2_2

$$w'(x,t) = 2cw(x,t) + \int_0^t N(t-s)\Delta w(x,s)ds + H(x,t), \qquad w(x,0) = u_0(x)$$

$$(2.3)$$

where

$$H(x,t) = \int_0^t F(x,s)ds + [v_0(x) - 2cu_0(x)], \quad N(t) = c_0^2 + \int_0^t M(s)ds. \quad (2.4)$$

We can pass from one system to the other and study these systems in a unified way, but they have different physical interpretations, see Sect. 2.6.

The "initial" time $t_0 = 0$ is the time after which a control f is applied to the system. Up to now, controllability has been mostly studied when the control acts in the Dirichlet boundary condition (the important case that the control is a traction on the boundary seems not yet sufficiently studied):

$$w(x,t) = f(x,t), \qquad x \in \Gamma \subseteq \partial\Omega, \qquad w(x,t) = 0, \qquad x \in \partial\Omega \setminus \Gamma. \quad (2.5)$$

We call Γ the ACTIVE PART of the boundary, and we do not exclude $\Gamma = \partial\Omega$. We study whether it is possible to force $(w(x,t), w_t(x,t))$ to hit prescribed targets at some time $T > 0$ (see the precise definition in Sect. 2.3).

These general statements are now sufficient for the introduction of the assumptions and suitable shorthand notations.

The following assumptions are always used, and not explicitly repeated:

- the region Ω (on one side of its boundary) is bounded with C^2 boundary.
- Γ is relatively open in $\partial\Omega$.
- the kernel $M(t)$ is of class $C^2(0,T)$ for every $T > 0$ and $c_0^2 > 0$. So, the kernel $N(t)$ is of class $C^3(0,T)$ for every $T > 0$ and $N(0) > 0$.

We shall see that $c_0^2 > 0$ implies that the signals propagate with finite velocity. The second principle of thermodynamics imposes further restrictions to the kernels $M(t)$ and $N(t)$ (see [33]), which are of no use in the study of controllability.

Notations
We recall that γ_0 denotes the trace on $\partial\Omega$ and γ_1 denotes the exterior normal derivative on $\partial\Omega$.

As stated in Sect. 1.3, the convolution is denoted $*$:

$$(f * g)(t) = \int_0^t f(t-s)g(s)ds.$$

Unless needed for clarity, dependence on the time or space variables is not indicated and the apex denotes time derivative. So, Eqs. (2.1) and (2.3) can be written respectively as

$$w'' = 2cw' + c_0^2 \Delta w + M * \Delta w + F , \qquad w' = 2cw + N * \Delta w + H .$$

The function f is called a *control* while the functions F and H are the (*distributed*) *affine terms*. As explained in Sect. 2.6, the functions F and H depend on the history of the system for $t < 0$.

Dependence of the solutions on the initial conditions and on the affine term is not indicated. We use w^f to denote dependence on the control f, when needed for clarity.

We have to consider both controlled systems, i.e., systems with f any (square integrable) function, and *uncontrolled systems*, i.e., systems with the Dirichlet boundary conditions put equal 0 on the whole of $\partial \Omega$. Solutions of controlled systems will be denoted u or w, while solutions of uncontrolled systems will be denoted with a Greek letter. For example, $\phi'' = \Delta \phi$, $\phi = 0$ on $\partial \Omega$.

Controllability is studied in *real* Hilbert spaces but, when needed the spaces are complexified without any specific observation (for example when defining the operator $\mathcal{A}$ and the sine and cosine operators).

The Semigroup and the Cosine Operator Generated by A
We consider the operator $A: L^2(\Omega) \mapsto L^2(\Omega)$ defined in (1.31):

$$\mathrm{dom}\, A = H^2(\Omega) \cap H_0^1(\Omega) , \qquad A\phi = \Delta \phi. \tag{2.6}$$

Some of its properties have been described in Sect. 1.3. In particular, we recall that $\{\phi_n\}$ denotes a sequence of eigenvectors of A, which is an orthormal basis of $L^2(\Omega)$.

We need further properties. First, we recall that a function $K(t)$ from $[a, b]$ to $\mathcal{L}(H)$ (H a Banach or Hilbert space) is a STRONGLY CONTINUOUS FUNCTION when the H-valued functions $K(t)h$ are continuous for every $h \in H$.

The operator valued function $t \mapsto e^{At}$ defined by

$$e^{A0} = I , \qquad e^{At}\left(\sum_{n=1}^{+\infty} \alpha_n \phi_n\right) = \sum_{n=1}^{+\infty} e^{-\lambda_n^2 t} \alpha_n \phi_n , \qquad t > 0 \tag{2.7}$$

is the STRONGLY CONTINUOUS SEMIGROUP generated by A. Its properties are:

- the operator e^{At} is defined for every $t \geq 0$ and $e^{At} \in \mathcal{L}(L^2(\Omega))$ is selfadjoint.
- for $t \geq 0$, $\tau \geq 0$ we have: $e^{A(t+\tau)} = e^{At}e^{A\tau}$ and $e^{A0} = I$.
- the transformation $t \mapsto e^{At}$ is strongly continuous.
- a strongly continuous semigroup is a STRONGLY CONTINUOUS GROUP when it is defined also for $t < 0$ and the properties stated above hold for t and τ in $\mathbb{R}$. It is known that the semigroup (2.7) *is not a group*.

Let $\mathscr{A} = i(-A)^{1/2}$. It is known (see [10, 27]) that $e^{\mathscr{A}t}$ is a C_0-*group* of operators on $L^2(\Omega)$. In terms of the Fourier expansion in (1.32) and (1.33) we have:

$$
\text{if} \quad y = \sum_{n=1}^{+\infty} \alpha_n \phi_n \in L^2(\Omega) \quad \text{then} \quad
\begin{cases}
\mathscr{A} y = i \left(\displaystyle\sum_{n=1}^{+\infty} \lambda_n \alpha_n \phi_n \right), \\
e^{\mathscr{A}t} y = \displaystyle\sum_{n=1}^{+\infty} e^{i\lambda_n t} \alpha_n \phi_n.
\end{cases}
\tag{2.8}
$$

The $\mathscr{L}(L^2(\Omega))$-operator valued function

$$
R_+(t) = \frac{1}{2} \left[e^{\mathscr{A}t} + e^{-\mathscr{A}t} \right] \qquad t \in \mathbb{R}
$$

is the COSINE OPERATOR generated by A. Its key property is the COSINE FORMULA

$$
R_+(t)R_+(\tau) = \frac{1}{2} \left(R_+(t) + R_+(\tau) \right) \qquad \forall t, \tau \in \mathbb{R}.
\tag{2.9}
$$

It is convenient to introduce the operators

$$
R_-(t) = \frac{1}{2} \left[e^{\mathscr{A}t} - e^{-\mathscr{A}t} \right], \qquad S(t) = \mathscr{A}^{-1} R_-(t), \qquad t \in \mathbb{R}.
$$

The operator $S(t)$ is the SINE OPERATOR (generated by A).

The following properties are known (see [27, 60, 91]):

- $R_+(t)$, $R_-(t)$ and $S(t)$ are selfadjoint continuous operators for every $t \in \mathbb{R}$ and they are strongly continuous functions of $t \in \mathbb{R}$.
- $S(t)$ takes values in dom $\mathscr{A}$.
- for all $z \in L^2(\Omega)$ we have $S(t)z = \int_0^t R_+(r)z\,dr$.
- for every $z \in$ dom $\mathscr{A}$ we have

$$
\frac{d}{dt} R_+(t)z = \mathscr{A} R_-(t)z = A S(t)z, \qquad \frac{d}{dt} R_-(t)z = \mathscr{A} R_+(t)z.
$$

- if $z \in \mathscr{D}(\Omega)$ then $R_+(t)z$, $R_-(t)z$, $e^{\mathscr{A}t}z$ and $e^{At}z$ are of class C^∞ because $\mathscr{D}(\Omega) \subseteq$ dom $\mathscr{A}^k$ for every k.
- The operators $R_+(t)$, $R_-(t)$, $S(t)$ transform $H_0^1(\Omega)$ to itself and can be extended by continuity to $H^{-1}(\Omega)$.

Formulas (2.8) give

$$
R_+(t) \left(\sum_{n=1}^{+\infty} \alpha_n \phi_n \right) = \sum_{n=1}^{+\infty} (\alpha_n \cos \lambda_n t) \, \phi_n(x) ,
$$

$$R_-(t)\left(\sum_{n=1}^{+\infty} \alpha_n \phi_n\right) = i\left(\sum_{n=1}^{+\infty} (\alpha_n \sin \lambda_n t)\, \phi_n(x)\right), \tag{2.10}$$

$$S(t)\left(\sum_{n=1}^{+\infty} \alpha_n \phi_n\right) = \left(\sum_{n=1}^{+\infty} \left(\alpha_n \frac{\sin \lambda_n t}{\lambda_n}\right) \phi_n(x)\right).$$

We shall use the following integration by parts formulas, which hold for $u(t) \in C^1(0, T; L^2(\Omega))$ (see [74]):

$$\int_0^t R_+(c_0(t-s))u'(s)ds = u(t) - R_+(c_0 t)u(0) + c_0 \mathcal{A} \int_0^t R_-(c_0(t-s))u(s)ds, \tag{2.11}$$

$$\int_0^t R_-(c_0(t-s))u'(s)ds = -R_-(c_0 t)u(0) + c_0 \mathcal{A} \int_0^t R_+(c_0(t-s))u(s)ds. \tag{2.12}$$

The Dirichlet Operator

We introduce the DIRICHLET OPERATOR D:

$$u = Df \iff \begin{cases} \Delta u(x) = 0 \text{ in } \Omega \\ u(x) = f(x) \text{ if } x \in \Gamma, \quad u(x) = 0 \text{ if } x \in \partial\Omega \setminus \Gamma. \end{cases} \tag{2.13}$$

The operator D, initially defined on "smooth" functions f, admits an extension $D \in \mathcal{L}\left(L^2(\Gamma), L^2(\Omega)\right)$. The function $u = Df \in L^2(\Omega)$ is the (unique) solution of the boundary value problem (2.13).

Let $\phi \in \mathrm{dom}\, A = H^2(\Omega) \cap H_0^1(\Omega)$. Then, we have (see [93, Proposition 10.6.1] and note that our operator A is $-A_0$ in [93]):

$$\int_\Omega A\phi Df\,dx = \int_\Gamma (\gamma_1\phi)\, f\,d\Gamma + \int_\Omega \phi\Delta Df = \int_\Gamma (\gamma_1\phi)\, f\,d\Gamma, \tag{2.14}$$

$$A\phi_n = -\lambda_n^2 \phi_n \implies \int_\Omega \phi_n Df\,dx = -\frac{1}{\lambda_n^2}\int_\Gamma (\gamma_1\phi_n)\, f\,d\Gamma. \tag{2.15}$$

2.1.1 The Wave Equation

We shall consider Eqs. (2.1) or (2.3) as a perturbation of a suitable wave type equation. With the goal of defining the solutions of the systems with persistent memory, we recall few crucial facts on the problem

$$u'' = c_0^2 \Delta u + G(x, t) , \quad \begin{cases} u(x, 0) = u_0(x), & u'(x, 0) = v_0(x), \\ u(x, t) = f(x, t) \; x \in \Gamma, \; u(x, t) = 0 \; x \in \partial\Omega \setminus \Gamma. \end{cases}$$
(2.16)

Thanks to the linearity of the problem, we can consider separately the dependence of the solution on the initial conditions (u_0, v_0), the distributed affine term G and the boundary control f. We have (see for example [60, 61, 66]):

Theorem 2.1 Let $T > 0$. The following properties hold for problem (2.16):

1. the transformation

$$(u_0, v_0) \mapsto \big(u(\cdot, t), u'(\cdot, t)\big) \; : \; H_0^1(\Omega) \times L^2(\Omega) \mapsto C([0, T]; H_0^1(\Omega) \times L^2(\Omega))$$

is (affine) linear and continuous. It is also (affine) linear and continuous as a transformation $L^2(\Omega) \times H^{-1}(\Omega) \mapsto C\big([0, T]; L^2(\Omega) \times H^{-1}(\Omega)\big)$. A short-hand notation is

$$u \in C([0, T]; L^2(\Omega)) \cap C^1([0, T]; H^{-1}(\Omega)) \, .$$

2. the transformation $G \mapsto \big(u(\cdot, t), u'(\cdot, t)\big)$ is (affine) linear and continuous from $L^1(0, T; L^2(\Omega))$ to $C\big([0, T]; H_0^1(\Omega) \times L^2(\Omega)\big)$, hence also from $L^1(0, T; L^2(\Omega))$ to $C\big([0, T]; L^2(\Omega) \times H^{-1}(\Omega)\big)$.
3. the transformation from the boundary control f to $\big(u(\cdot, t), u'(\cdot, t)\big)$ is (affine) linear and continuous from $L^2(0, T; L^2(\partial\Omega))$ to $C\big([0, T]; L^2(\Omega) \times H^{-1}(\Omega)\big)$.

So, (when the control acts on the boundary, the case we shall study) controllability has to be considered in $L^2(\Omega) \times H^{-1}(\Omega)$.

The solution $u = u(x, t) \in C([0, T]; L^2(\Omega)) \cap C^1([0, T]; H^{-1}(\Omega))$ of Problem (2.16) is given by (see [60, 61] and recall $c_0 > 0$)

$$u(t) = R_+(c_0 t)u_0 + \frac{1}{c_0}\mathscr{A}^{-1}R_-(c_0 t)v_0 + \frac{1}{c_0}\mathscr{A}^{-1}\int_0^t R_-(c_0(t - s))G(s)ds$$

$$- c_0\mathscr{A}\int_0^t R_-(c_0(t - s))Df(s)ds,$$
(2.17)

$$u'(t) = c_0 \mathscr{A} R_-(c_0 t)u_0 + R_+(c_0 t)v_0 + \int_0^t R_+(c_0(t-s))G(s)ds$$

$$- c_0^2 A \int_0^t R_+(c_0(t-s))Df(s)ds. \tag{2.18}$$

The function u in (2.17) is not a differentiable function, and cannot be replaced in both the sides of the equation. So, we define:

Definition 2.1 The function u in (2.17) is the MILD SOLUTION of Eq. (2.16) (in the next chapters the term "solution" will always denote a mild solution). It is a REGULAR SOLUTION when

1. $u(t) - Df(t) \in C([0, T]; \operatorname{dom} A) \cap C^1([0, T]; \operatorname{dom}\mathscr{A}) \cap C^2([0, T]; L^2(\Omega))$.
2. $u(0) = u_0$ and $u'(0) = v_0$.

Note the sense in which the boundary condition is satisfied by the regular solutions:

$$u(t) - Df(t) \in \operatorname{dom} A = H^2(\Omega) \cap H_0^1(\Omega). \tag{2.19}$$

The next result will be used to justify the definition of mild solution:

Theorem 2.2 *Let* $u_0 \in \operatorname{dom} A$, $v_0 \in \operatorname{dom} \mathscr{A}$, $G \in \mathscr{D}(\Omega \times (0, T))$ *and* $f \in \mathscr{D}(\Gamma \times (0, T))$. *Then, the function* $u(t)$ *in* (2.17) *is a regular solution of Eq.* (2.16) *and the following equality holds for every* t:

$$u''(t) = c_0^2 A(u(t) - Df(t)) + G(t). \tag{2.20}$$

Proof We introduce the functions

$$u_1(t) = R_+(c_0 t)u_0 + \frac{1}{c_0}\mathscr{A}^{-1} R_-(c_0 t)v_0 ,$$

$$u_2(t) = \frac{1}{c_0}\mathscr{A}^{-1} \int_0^t R_-(c_0(t-s))G(s)ds = \frac{1}{c_0}\mathscr{A}^{-1} \int_0^t R_-(c_0 s)G(t-s)ds ,$$

$$u_3(t) = -c_0\mathscr{A} \int_0^t R_-(c_0(t-s))Df(s)ds = -c_0\mathscr{A} \int_0^t R_-(c_0 s)Df(t-s)ds .$$

Thanks to the linearity we examine separately these functions.

The definitions of $R_+(t)$ and $R_-(t)$ clearly imply that $u_1(t)$ is a regular solution of (2.20) with $f = 0$, $G = 0$.

We consider the function $u_3(t)$ and we leave the similar analysis of $u_2(t)$ to the reader. Let $y(t) = u_3(t) - Df(t)$. First we note that

$$y(t) = -c_0 \mathscr{A} \int_0^t R_-(c_0 s) Df(t-s)ds - Df(t) \,,$$

$$y'(t) = -c_0 \mathscr{A} \int_0^t R_-(c_0 s) Df'(t-s)ds - Df'(t) \,,$$

$$y''(t) = -c_0 \mathscr{A} \int_0^t R_-(c_0 s) Df''(t-s)ds - Df''(t) \,.$$

An integration by parts, using formula (2.11), proves $y''(t) \in C([0, T]; L^2(\Omega))$. In fact,

$$y''(t) = -\int_0^t R_+(c_0 s) Df'''(t-s)ds \,. \tag{2.21}$$

Integrating by parts twice we get $y' \in C([0, T]; \mathrm{dom}\mathscr{A})$. In fact:

$$y'(t) = -\frac{1}{c_0} \mathscr{A}^{-1} \int_0^t R_-(c_0 s) Df'''(t-s)ds.$$

We give the details of the proof that $y(t) \in C([0, T]; \mathrm{dom} A)$. We integrate by parts three times, using both the formulas (2.11) and (2.12):

$$y(t) = -\int_0^t R_+(c_0 s) Df'(t-s)ds = -\frac{1}{c_0} \mathscr{A}^{-1} \int_0^t R_-(c_0 s) Df''(t-s)ds$$

$$= \frac{1}{c_0^2} A^{-1} \left[Df''(t) - \int_0^t R_+(c_0 s) Df'''(t-s)ds \right].$$

We compare this last equality and (2.21), using $y = u_3 - Df$, and we see that u_3 solves (2.20) (with $G = 0$):

$$c_0^2 Ay(t) = Df''(t) - \int_0^t R_+(c_0 s) Df'''(t-s)ds = Df''(t) + y''(t) = u_3''(t) \,.$$

The linearity of the problem shows that $u(t) = u_1(t) + u_2(t) + u_3(t)$ solves (2.20), hence also (2.16) since when $\phi \in \mathrm{dom} A$, then $A\phi = \Delta\phi$.

Remark 2.1 Stronger properties give more regular solutions. Note that $\mathscr{D}(\Omega) \subseteq$ dom A^k for every k. So, if $f = 0$, $G = 0$ and u_0, v_0 belong to $\mathscr{D}(\Omega)$, then $u(t) \in C^\infty([0, T]; H^r(\Omega))$ for every r and so $u(t) \in C^\infty(\Omega \times (0, T))$ and $\gamma_0 u(t)$, $\gamma_1 u(t)$ belong to $C^\infty([0, T]; L^2(\partial\Omega))$. In particular, if $D_i = \partial/\partial x_i$ then $D_i u(x, t)$ solves (2.16) with initial conditions $D_i u_0$ and $D_i v_0$ (but in general $\gamma_0 D_i u \neq 0$). If furthermore $\partial\Omega \in C^\infty$, then the derivatives of u of every order have smooth traces on $\partial\Omega$.

Different methods can be used to justify the definition of mild solutions, see [27, 84]. We adopt the following one: let u_n be given by (2.17) with data $\{f_n\}$, $\{G_n\}$, $\{u_{0,n}\}$, $\{v_{0,n}\}$ each one of class C^∞ with compact support and such that

$$u_{0,n} \to u_0 \text{ in } L^2(\Omega), \quad v_{0,n} \to v_0 \text{ in } H^{-1}(\Omega),$$
$$f_n \to f \text{ in } L^2(0, T; L^2(\partial\Omega)), \quad G_n \to G \text{ in } L^2(0, T; L^2(\Omega)).$$

Then, u_n is a regular solution and Theorem 2.1 show that

$$u_n \to u \text{ in } C([0, T]; L^2(\Omega)), \quad u_n' \to u' \text{ in } C([0, T]; H^{-1}(\Omega))$$

for every $T > 0$: *a mild solution is the limit of a sequence of regular solutions.* This observation justifies the definition of mild solution.

Remark 2.2 The property in Item 3 of Theorem 2.1 is called ADMISSIBILITY of the operator $\begin{bmatrix} 0 \\ c_0^2 AD \end{bmatrix}$.

2.2 The Solutions of the System with Memory

The properties of the wave equation just outlined can be used to derive definitions and formulas for the solutions of the systems with memory (2.1) and (2.3). We must find a suitable formula, which can be used to *define* the solution, and then of course we must justify our choice. With this goal in mind, we perform formal manipulations as follows. First a transformation, which is known as MACCAMY TRICK. We rewrite the Eq. (2.1) in the form

$$\Delta w + \frac{1}{c_0^2} M * \Delta w = \frac{1}{c_0^2} w'' - \frac{2c}{c_0^2} w' - \frac{1}{c_0^2} F \tag{2.22}$$

and we denote $R(t)$ the resolvent kernel of $M(t)/c_0^2$ so that (see Sect. 1.3)

$$R = \frac{1}{c_0^2} M - \frac{1}{c_0^2} M * R. \tag{2.23}$$

Then we have

$$c_0^2 \Delta w = w'' - 2cw' - F - R * w'' + 2cR * w' + R * F .$$

We integrate by parts the integrals $R * w''$ and $R * w'$ and we get:

$$w''(t) = c_0^2 \Delta w(t) + aw' + bw + \int_0^t K(t-s)w(s)ds + F_1(t), \qquad (2.24)$$

$$F_1(t) = F(t) - \int_0^t R(t-s)F(s)ds - R(t)v_0 - (R'(t) - 2cR(t))u_0 ,$$

$$a = 2c + R(0) , \quad b = R'(0) - 2cR(0) , \quad K(t) = \left[R''(t) - 2cR'(t) \right].$$

The kernel $K(t)$ is continuous.

Remark 2.3 The noticeable fact of the MacCamy trick is that *the laplacian does not appear in the memory term of* (2.24). Furthermore, note that when $a = 0$, $b = 0$ we get a system a special case of which has been studied in Chap. 1 (when the space variable is in $(0, +\infty)$).

Equation (2.24) is the same as the wave equation (2.16) with $G(t) = F_1(t) + aw' + bw + K * w$. So, we can combine formulas (2.17) and (2.24) to obtain a Volterra integral equation for $w(t)$:

$$w(t) = -c_0 \mathscr{A} \int_0^t R_-(c_0(t-s))Df(s)ds + \frac{1}{c_0}\mathscr{A}^{-1}\int_0^t R_-(c_0(t-s))F_1(s)ds$$

$$+ R_+(c_0t)u_0 + \frac{1}{c_0}\mathscr{A}^{-1}R_-(c_0t)v_0 + \frac{a}{c_0}\mathscr{A}^{-1}\int_0^t R_-(c_0(t-s))w'(s)ds$$

$$+ \frac{1}{c_0}\mathscr{A}^{-1}\int_0^t R_-(c_0(t-s))\left[bw(s) + \int_0^s K(s-r)w(r)dr \right] ds$$

$$= H(t) + \int_0^t I(s)w(t-s)ds. \qquad (2.25)$$

where (use (2.12) to integrate by parts the integral which contains w')

$$H(t) = -c_0 \mathscr{A} \int_0^t R_-(c_0(t-s))Df(s)ds + \frac{1}{c_0}\mathscr{A}^{-1} \int_0^t R_-(c_0(t-s))F_1(s)ds$$

$$+ R_+(c_0 t)u_0 + \frac{1}{c_0}\mathscr{A}^{-1}R_-(c_0 t)\,[v_0 - au_0]\ ,$$

$$L(t)w = \left[aR_+(c_0 t) + \frac{b}{c_0}\mathscr{A}^{-1}R_-(c_0 t) \right] w + \frac{1}{c_0}\mathscr{A}^{-1} \int_0^t K(r)R_-(c_0(t-r))wdr\ .$$

The affine term $H(t)$ appears in the solutions of the wave equation (2.17) and so its properties are known (see Theorem 2.1 and recall that the function $F_1(t)$ here depends also on u_0 and v_0): $H(t) \in C([0, T]; L^2(\Omega)) \cap C^1([0, T]; H^{-1}(\Omega))$. The properties of $L(t)$ are:

- $L(t) \in \mathscr{L}(L^2(\Omega))$ for every $t \geq 0$ and $t \mapsto L(t)$ is strongly continuous.
- $L(t)L(\tau) = L(\tau)L(t)$ for every $t \geq 0,\ \tau \geq 0$.

So, using known properties of the Volterra integral equations (see Sect. 1.3):

Theorem 2.3 *Let $T > 0$. There exists a unique solution $w \in C([0, T]; L^2(\Omega))$ with $w' \in C([0, T]; H^{-1}(\Omega))$ of the Volterra integral equation (2.25) which depends continuously on $u_0 \in L^2(\Omega)$, $v_0 \in H^{-1}(\Omega)$, $F \in L^1(0, T; L^2(\Omega))$ and $f \in L^2(0, T; L^2(\Gamma))$.*

We use the Volterra integral Eq. (2.25) to define the solutions of Eq. (2.1):

Definition 2.2 The MILD SOLUTION of (2.1) (or of (2.3)) with conditions (2.2) and (2.5) is the function $w(t)$ which solves (2.25). The mild solution is a REGULAR SOLUTION when

$$w(t) - Df(t) \in C([0, T]; \mathrm{dom}\,A) \cap C^1([0, T]; \mathrm{dom}\,\mathscr{A}) \cap C^2([0, T]; L^2(\Omega))\ .$$

In the next chapters, the term "solution" will always denote a mild solution.

Also in the case of systems with memory, regular solutions satisfy the boundary conditions in the sense that $w(t) - Df(t) \in \mathrm{dom}\,A = H^2(\Omega) \cap H_0^1(\Omega)$.

Remark 2.4 Note that $w(t)$ takes real values when the initial conditions, the affine term and the control are real.

The following result justifies the definition of the regular and mild solutions. The first statement is a reformulation of Theorem 2.3.

Theorem 2.4 *Let $T > 0$ be fixed.*

1. *The transformation $(u_0, u_1, F, f) \mapsto w$ is continuous from $L^2(\Omega) \times H^{-1}(\Omega) \times L^1(0, T; L^2(\Omega)) \times L^2(0, T; L^2(\Gamma))$ to $C([0, T]; L^2(\Omega)) \cap C^1([0, T]; H^{-1}(\Omega))$.*

2. *Let $u_0 \in \mathrm{dom}\,A$, $v_0 \in \mathrm{dom}\,\mathscr{A}$, $F \in \mathscr{D}(\Omega \times (0, T))$ and $f \in \mathscr{D}(\Gamma \times (0, T))$.*
   Then, the mild solution defined by the Volterra integral equation (2.25) is a regular
   solution and the following equalities hold for every t (here $y(t) = w(t) - Df(t)$):

$$w''(t) = c_0^2 Ay(t) + aw'(t) + bw(t) + \int_0^t K(t - s)w(s)ds + F_1(t), \quad (2.26)$$

$$w''(t) = 2cw' + c_0^2 Ay(t) + \int_0^t M(t - s)Ay(s)ds + F(t). \quad (2.27)$$

Proof We prove the second statement. Thanks to the linearity of the problem, as in
the case of the memoryless wave equation, we can study separately the effect of the
initial conditions, of the affine term and of the control. We confine ourselves to study
the effect of the control. So, $u_0 = 0$, $v_0 = 0$ and $F = 0$. We use the function u_3
introduced in the proof of Theorem 2.2 but now $y(t) = w(t) - Df(t)$. With these
notations, Eq. (2.25) takes the form

$$y(t) = (u_3(t) - Df(t)) + \int_0^t L(s)Df(t - s)ds + \int_0^t L(s)y(t - s)ds. \quad (2.28)$$

Using $f \in \mathscr{D}(\Gamma \times (0, T))$ and Theorem 2.2 we know that

$$A\,(u_3 - Df), \quad \mathscr{A}\,(u_3 - Df)', \quad (u_3 - Df)'' \text{ belong to } C([0, T]; L^2(\Omega)). \quad (2.29)$$

We prove the analogous properties of $y = w - Df$.
   Using the definition of $L(t)$ and Theorem 2.2 we can prove:

$$A \int_0^t L(t - s)Df(s)ds \in C([0, T]; L^2(\Omega)),$$
$$\mathscr{A} \int_0^t L(t - s)Df'(s)ds \in C([0, T]; L^2(\Omega)). \quad (2.30)$$

We sketch the proof of the first property (the proof of the second one is similar).
The definition of $L(t)$ shows that $A \int_0^t L(t - s)Df(s)ds$ is the linear combination of
three terms. It is easy to see that they belong to $C([0, T]; L^2(\Omega))$. In fact, the first
one is (integrate by parts twice)

$$A \int_0^t R_+(c_0(t-s))Df(s)ds = -\frac{1}{c_0^2}\left[Df'(t) - \int_0^t R_+(c_0(t-s))Df''(s)ds\right],$$

an $L^2(\Omega)$-valued continuous function. The second and third terms are treated analogously.

Now, we proceed to prove that $w(t)$ is a regular solution. We first prove that $y = w - Df \in C([0, T]; \mathrm{dom}\,A)$. This follows because (2.28) shows that $Ay(t)$ solves

$$Ay(t) = A\left(u_3(t) - Df(t) + \int_0^t L(t-s)Df(s)ds\right) + \int_0^t L(t-s)Ay(s)ds$$

so that $Ay \in C([0, T]; L^2(\Omega))$ i.e., $y \in C([0, T]; \mathrm{dom}\,A)$ thanks to the first properties in (2.29) and in (2.30).

Now, we study the derivatives of y. The equations of $\xi = y'$ and $\zeta = y''$ are respectively

$$\xi = (u_3 - Df)' + \int_0^t L(s)Df'(t-s)ds + \int_0^t L(s)\xi(t-s)ds, \qquad (2.31)$$

$$\zeta = (u_3 - Df)'' + \int_0^t L(s)Df''(t-s)ds + \int_0^t L(s)\zeta(t-s)ds \qquad (2.32)$$

(note that $y(0) = 0$, $y'(0) = 0$). From (2.32) and continuity of $(u_3 - Df)''$ we get $\zeta \in C([0, T]; L^2(\Omega))$. We consider Eq. (2.31): the second property in (2.30) and continuity of $\mathscr{A}(u_3 - Df)'$ prove $\mathscr{A}\xi \in C([0, T]; L^2(\Omega))$, as wanted.

Now, we prove the equalities (2.26) and (2.27). We present the proof in the case $u_0 = 0$, $v_0 = 0$ and $F = 0$. We proved that w is twice differentiable (when $f \in \mathscr{D}(\Gamma \times (0, T))$) so that we can integrate by parts back in (2.25). Then, (2.25) takes the form

$$w(t) = -c_0\mathscr{A}\int_0^t R_-(c_0(t-s))Df(s)ds$$

$$+ \frac{1}{c_0}\mathscr{A}^{-1}\int_0^t R_-(c_0(t-s))\left[aw'(s) + bw(s) + \int_0^s K(s-r)w(r)dr\right]ds .$$

We compare with the functions $u_2(t)$ and $u_3(t)$ in Theorem 2.2 and with (2.20) and we see that (2.26) holds.

Using again differentiability of $w(t)$ and the definition of $K(t)$, we integrate by parts the last integral in (2.26). We find:

$$(w'' - 2cw') - R * (w'' - 2cw') = c_0^2 Ay. \tag{2.33}$$

The definition (2.23) of $R(t)$ gives

$$M * Ay = R * (w'' - 2cw') .$$

Using (2.33) to replace the convolution on the right side, we find (2.27).

We sum up: *the definition of the mild solutions is justified since mild solutions are limits of regular solutions.*

Remark 2.5 It is easy to extend Remark 2.1 to Eq. (2.26): if $f = 0$, $F = 0$ and both the initial data u_0 and v_0 belong to $\mathscr{D}(\Omega)$ then $w(t) \in C^\infty([0, T]; H^r(\Omega))$ for every r and so it has continuous derivatives of every order. In particular, $D_i w(x, t)$ solves Eq. (2.26) with initial conditions $D_i u_0$ and $D_i v_0$ ($D_i w(x, t)$ might not be zero on the boundary). If furthermore $\partial\Omega \in C^\infty$ then the derivatives of w of every order have smooth traces on $\partial\Omega$.

2.3 Description of the Control Problems

Now, we describe the control problem. In the study of linear systems, exact or approximate controllability does not depend on the choice of initial conditions and affine terms (which are kept fixed). For this reason *from now on, when studying control problems, we assume*

$$u_0 = 0 , \quad v_0 = 0 , \quad F = 0 .$$

Definition 2.3 A "target" $(\xi, \eta) \in L^2(\Omega) \times H^{-1}(\Omega)$ is *reachable at time T* when there exists a STEERING CONTROL $f \in L^2(0, T; L^2(\Gamma))$ such that

$$w^f(T) = \xi , \quad (w^f)'(T) = \eta .$$

The REACHABLE SET at time T is the set of the reachable targets (at time T) and system (2.1) is CONTROLLABLE (at time T) when *every* target in $L^2(\Omega) \times H^{-1}(\Omega)$ is reachable (APPROXIMATE CONTROLLABILITY when the reachable set is dense in $L^2(\Omega) \times H^{-1}(\Omega)$. In order to contrast approximate controllability and controllability, the last property is also called EXACT CONTROLLABILITY).

Note that:

- targets are taken to be real so that we use *real control functions.*
- if a target is reachable at time T_0 it is also reachable at any later time $T > T_0$, using the control $f(t - (T - T_0))$ (equal to zero if $t < T - T_0$).
- we might wish to study controllability of the sole component w. This has an interest in some applications, see the identification problem in Sect. 5.4.

It might seem that we can relax the property that the control time is "universal", and that we might require that every target be reached in a certain time T, not the same for all of them. In fact, this would not give a more general definition of controllability, since we can prove:

Theorem 2.5 *Assume that for every target $\xi \in L^2(\Omega)$ there exists a time $T = T_\xi$ and a control f such that $w(T_\xi) = \xi$. Then, there exists a time T_0 such that the system is controllable at time T_0.*
A similar statement holds for the pair (w, w')

Proof In this proof we use the following facts (see [11, Chap. 3]. The statements are adapted to Hilbert spaces): (1) a *convex* set is closed in the norm topology if and only if it is weakly closed, i.e., if every weakly convergent sequence converges to a point of the set; (2) a *bounded* set which is *convex* and *closed* is also weakly compact, i.e., every sequence in the set admits a weakly convergent subsequence, which converges to a point of the set; (3) any linear and continuous transformation is weakly continuous, and so it transforms weakly compact sets into weakly compact sets. Furthermore, we shall use Baire Theorem, see below.

Now we prove the theorem, studying controllability of the sole component w. The proof of the second assertion, concerning the pair (w, w'), is similar. Let

$$R(T) = \left\{ w^f(T), \quad f \in L^2(0, T; L^2(\Gamma)) \right\},$$
$$R_N(T) = \left\{ w^f(T), \quad |f|_{L^2(0,T;L^2(\Gamma))} \leq N \right\}.$$

By assumption, every target can be reached. So,

$$L^2(\Omega) = \bigcup_{T>0,\, N\in\mathbb{N}} R_N(T) = \bigcup_{T\in\mathbb{N},\, N\in\mathbb{N}} R_N(T)$$

(the first equality is the assumption of the theorem and the second follows since $T \mapsto R_N(T)$ and $N \mapsto R_N(T)$ are increasing).
The set

$$\left\{ f, \quad |f|_{L^2(0,T;L^2(\Gamma))} \leq N \right\}$$

is convex and weakly compact in $L^2(0, T; L^2(\Gamma))$ and $R_N(T) \subseteq L^2(\Omega)$ is the image of such convex weakly compact set under a linear continuous transformation.

Hence $R_N(T)$ is convex and weakly compact, hence also weakly closed. So, it is closed in the norm topology of $L^2(\Omega)$.

We sum up: the space $L^2(\Omega)$ is union of the (double) sequence $\{R_N(T)\}_{N,T\in\mathbb{N}}$ of *closed sets*. Baire Theorem (see [72, 120]) implies the existence of T_0 and N_0 such that $R_{N_0}(T_0)$ has a nonempty interior. Hence, also

$$R(T_0) = \bigcup_{N\in\mathbb{N}} R_N(T_0) \subseteq L^2(\Omega)$$

is a subspace with a nonempty interior. Hence, it must be the whole space $L^2(\Omega)$ and *every* $\xi \in L^2(\Omega)$ is reachable in time T_0.

Remark 2.6 In the memoryless case and for distributed control the analogous of Theorem 2.5 has been first proved in [30] and then independently reproved or extended by several authors, essentially with the same proof, see for example [73, 83, 86].

The content of Theorem 2.5 combined with the fact that a system, which is controllable at a certain time T is also controllable at later times justifies the following definition:

Definition 2.4 The infimum of those times T at which the system is controllable is called the SHARP CONTROL TIME.

2.4 Useful Transformations

In this section, we show different representations of the control systems (2.1) or (2.3), or of their solutions, *which do not change the controllability property*. Different approaches to controllability can profit of one or the other representation. Accepting these facts, the proofs in this section can be skipped. However, the reader should keep in mind that the transformation used to achieve the condition $c_0 = 1$, i.e., $N(0) = 1$, changes the control time.

1. Let us represent

$$w(x, t) = \sum_{n=1}^{+\infty} \phi_n(x)w_n(t) , \qquad w_n(t) = \int_\Omega \phi_n(x)w_n(x, t)\mathrm{d}x \qquad (2.34)$$

($\{\phi_n\}$ orthonormal basis of real eigenvectors of A in $L^2(\Omega)$). It turns out that $w_n(t)$ satisfies

$$w_n''(t) = 2cw_n'(t) - \lambda_n^2\left(c_0^2 w_n(t) + \int_0^t M(t-s)w_n(s)\mathrm{d}s\right)$$

$$-\int_{\Gamma} \left(c_0^2 f(t) + \int_0^t M(t-s) f(s) ds \right) (\gamma_1 \phi_n) d\Gamma + F_n(t) \qquad (2.35)$$

$$w_n(0) = u_{0,n} = \int_{\Omega} u_0(x) \phi_n(x) dx, \quad w_n'(0) = v_{0,n} = \int_{\Omega} v_0(x) \phi_n(x) dx,$$

$$F_n(t) = \int_{\Omega} F(x,t) \phi_n(x) dx.$$

We integrate both the sides and we get:

$$w_n'(t) = 2c w_n(t) - \lambda_n^2 \int_0^t N(t-s) w_n(s) ds$$

$$-\int_0^t N(t-s) \left(\int_{\Gamma} (\gamma_1 \phi_n) f(x,s) d\Gamma \right) ds + H_n(t), \qquad (2.36)$$

$$w_n(0) = u_{0,n}, \qquad H_n(t) = \int_{\Omega} H(x,t) \phi_n(x) dx$$

(H defined in (2.4)) and so

$$w_n(t) = e^{2ct} u_{0,n} - \int_0^t e^{2c(t-\tau)} \left\{ \lambda_n^2 \int_0^\tau N(\tau-s) w_n(s) ds \right.$$

$$\left. + \int_0^\tau N(\tau-s) \left(\int_{\Gamma} (\gamma_1 \phi_n) f(x,s) d\Gamma \right) ds + H_n(\tau) \right\} d\tau. \qquad (2.37)$$

2. It is possible to remove the velocity term from formula (2.24). Let

$$w_\alpha(t) = e^{-\alpha t} w(t), \qquad \alpha = \frac{a}{2}.$$

Then, $w_\alpha(t)$ solves

$$\begin{cases} w_\alpha'' = c_0^2 \Delta w_\alpha + h w_\alpha + \int_0^t K_\alpha(t-s) w_\alpha(s) ds + F_\alpha(t), \\[2mm] w_\alpha(0) = u_0, \qquad w_\alpha'(0) = v_0 - \alpha u_0, \\ w_\alpha(t) = e^{-\alpha t} f(t) \text{ on } \Gamma, \quad w_\alpha(t) = 0 \text{ on } \partial\Omega \setminus \Gamma, \\ h = (b - \alpha^2), \qquad F_\alpha(t) = e^{-\alpha t} F(t), \quad K_\alpha(t) = e^{-\alpha t} K(t). \end{cases} \qquad (2.38)$$

3. When we study the control system represented in the form (2.3), it is not restrictive
to assume

$$c_0^2 = N(0) = 1 , \quad N'(0) = 0. \tag{2.39}$$

Remark 2.7 It is possible to study the convergence of the series (2.34) using the
formulas for $w_n(t)$ (see [77]). This is a different way to define the solutions of
system (2.1) and to study their properties.

Justification of the Formulas

Formal computations can easily justify the previous formulas. For example, recalling
that the eigenfunctions $\phi_n(x)$ are real, equality (2.35) is formally obtained computing
the inner product of both the sides of (2.1) with $\phi_n(x)$ and using formula (2.15). The
rigorous justifications are as follows:

1. The easiest way to prove formulas (2.35) and (2.36) is to show correctness for
regular solutions, and then to extend by continuity. So, let u_0 and v_0 belong to $\mathcal{D}(\Omega)$,
$F \in \mathcal{D}(\Omega \times (0, T))$ and $f \in \mathcal{D}(\Gamma \times (0, T))$. We integrate the product of both the
sides of (2.27) with ϕ_n and we use formula (2.15). Formulas (2.35), i.e., (2.36)
and (2.37), are easily derived under these special assumptions. We can pass to the
limit in (2.37), thanks to the first assertion of Theorem 2.4 and continuity of the inner
product, and so we conclude that formula (2.37) holds in general. Hence, also the
formulas (2.35) and (2.36) hold.

2. The series exapansion of $e^{-\alpha t}w(x, t)$ has coefficients $w_{\alpha,n}(t)$ solutions of an
equation similar to (2.35), but with $M(t)$ and $F(t)$ replaced with $F_\alpha(t)$ and $M_\alpha(t)$;
the initial conditions replaced by the Fourier coefficients of u_0 and $v_0 - \alpha u_0$ and
$e^{-\alpha t} f(t)$ in the place of $f(t)$. These are the coefficients of the Fourier expansion of
the solutions of (2.38).

3. Finally, we show how the conditions (2.39) can be achieved.

We use the representation (2.34) of $w(t)$ and the Eq. (2.36) for $w_n(t)$. The condition
$c_0^2 = N(0) = 1$ is achieved using the transformation

$$w \mapsto \theta = \sum_{n=1}^{+\infty} \phi_n(x)\theta_n(t) , \qquad \theta_n(x, t) = w_n(t/c_0) , \quad c_0 = \sqrt{N(0)} > 0 .$$

In fact,

$$\theta_n'(t) = 2\frac{c}{c_0}\theta_n - \lambda_n^2 \int_0^t \frac{1}{c_0^2} N((t-s)/c_0) \left[\theta_n(s) + \int_\Gamma (\gamma_1\phi_n) f(s/c_0)d\Gamma \right] ds$$

$$+ \frac{1}{c_0} H_n(t/c_0) .$$

Clearly, controllability of w is equivalent to controllability of θ while controllability of (w, w') is equivalent to controllability of (θ, θ'), but the control time is
divided by c_0: *if θ is controllable at the time S then w is controllable at the time*

$T = S/c_0 = S/\sqrt{N(0)}$ (and conversely). The bonus of this transformation is that

$$\text{the kernel is } N_1(t) = \frac{1}{c_0^2} N(t/c_0) \text{ so that} N_1(0) = 1 .$$

A second transformation is as follows: we introduce

$$\tilde{\theta}(x, t) = e^{2\gamma t}\theta(x, t) , \text{ i.e., } \tilde{\theta}_n(x, t) = e^{2\gamma t}\theta_n(x, t) , \quad \gamma = -\frac{1}{2}N_1'(0) .$$

Then, we have, with $N_2(t) = e^{2\gamma t} N_1(t)$ and $\alpha = (\gamma + c/c_0)$

$$\tilde{\theta}_n'(t) = 2\alpha\tilde{\theta}_n(t) - \lambda_n^2 \int_0^t N_2(t-s) \left[\tilde{\theta}_n(s) + \int_\Gamma (\gamma_1\phi_n) \left(e^{2\gamma s} f(s/c_0)\right) d\Gamma \right] ds$$

$$+ \frac{1}{c_0}e^{2\gamma t} H_n(t/c_0). \tag{2.40}$$

Neither controllability nor the control time are affected and

$$N_2(0) = 1 , \quad N_2'(0) = 2\gamma N_1(0) + N_1'(0) = 0 .$$

2.4.1 Finite Propagation Speed

FINITE PROPAGATION SPEED of the associated wave equation (2.16) is the following property (see for example [66, Remarque 1.2]): let $G = 0$ and let u_0 and v_0 have support in a compact set $K \subseteq \Omega$. Let $d > 0$ be *smaller then* the distance of x_0 from $K \cup \Gamma$. The known property is that[2] if $c_0 t < d$ then $u(x, t) = 0$ in $B(x_0, d - c_0 t)$. This can be expressed as follows: let $S(t) \subseteq \Omega$ be the union of the supports of the three functions (of the space variable x, while the time t is fixed)

$$R_+(c_0t)u_0 , \quad \mathscr{A}^{-1}R_-(c_0t)v_0 , \quad u(t) = -\mathscr{A}\int_0^t R_-(c_0(t-s))Df(s)ds. \tag{2.41}$$

Then,

$$c_0 t < d \implies S(t) \cap B(x_0, d - c_0 t) = \emptyset. \tag{2.42}$$

We prove that this property is retained by Eq. (2.1). We are interested in the propagation of signals produced by the control so we confine ourselves to the case

---

[2] $B(x_0, r)$ denotes the ball of center x_0 and radius r.

$u_0 = 0$, $v_0 = 0$ and $F = 0$. Furthermore, we can assume that f is "smooth", $f \in \mathscr{D}(\Gamma \times (0, +\infty))$.

It is easier if we represent Eq. (2.1) in the form (2.38) (index α is not indicated) so that (the definition of $u(t)$ is in (2.41))

$$
w(t) = u(t) + \frac{1}{c_0} \mathscr{A}^{-1} \int_0^t R_-(c_0(t-s)) \left[hw(s) + \int_0^s K(s-r)w(r)dr \right] ds. \quad (2.43)
$$

The solution of this Volterra integral equation is represented using the Picard series (1.20).

Our goal is the proof that the support of $w(x, t)$ does not intersect $B(x_0, d - c_0 t)$ if $c_0 t < d$. It is sufficient to prove this property for every term of the Picard series. In fact, it is enough to see this fact for the first term of the series, since it is clear that the method can be applied to every term.

The first term of the series is

$$
\frac{h}{c_0} \int_0^t \mathscr{A}^{-1} R_-(c_0(t-s)) u(s) ds \quad (2.44)
$$

$$
+ \frac{1}{c_0} \int_0^t \mathscr{A}^{-1} R_-(c_0(t-s)) \int_0^s K(s-r) u(r) dr \, ds. \quad (2.45)
$$

We consider the addendum (2.44). We already noted that the support of $u(t)$ does not intersect $B(x_0, d - c_0 t)$. Then, with s in the place of t, the support of $u(s)$ does not intersect $B(x_0, d - c_0 s)$. Hence,

the support of $\mathscr{A}^{-1} R_-(c_0(t-s)) u(s)$ does not intersect $B(x_0, d - c_0 t)$.

This property is retained by the integral on $[0, t]$.

The integral (2.45) is treated analogously and so the support of the first term in the Picard series does not intersect $B(x_0, d - c_0 t)$. It is clear that this property holds for every term of the series. Hence, we have:

Theorem 2.6 *If the distance of x_0 from Γ is larger than d and if $c_0 t < d$, then the support of $w(t)$ does not intersect $B(x_0, d - c_0 t)$.*

The interpretation is that signals (the "waves" in the body) propagate in a viscoelastic material with finite velocity not larger then c_0 (strict positivity of c_0^2 is crucial for this result). This observation suggests the following result:

Theorem 2.7 *The sharp control time of the system with memory is not shorter than that of the associated wave equation.*

We refer to [82] for the proof.

Remark 2.8 We mention that the velocity of the signals in the viscoelastic body is precisely c_0, the same as for the memoryless wave equation, hence *independent on* c and $M(t)$, see [22, 28, 46].

2.5 Final Comments

The solutions of the Eqs. (2.1) and (2.3) have been defined in this chapter using an "operator" approach as in [10, 74]. Fourier expansions can be used, see Remark 2.7. A different approach, the definition "by transposition" is related to the arguments in Chap. 6. The solutions can also be defined using the semigroup approach introduced in [18, 19]. This approach has a particular interest in the study of stability, and we cite the book [67].

A formal integration of both the sides of Eq. (2.3) gives

$$w(t) = u_0 + 2c \int_0^t w(s) ds + \int_0^t \left[\int_0^{t-r} N(s) ds \right] \Delta w(r) dr + \int_0^t F(s) ds \ .$$

In this form, the equation is deeply studied in [85], using Laplace transform techniques. See [40] for control problems.

An important case in applications is when the kernel is a linear combination of exponentials,

$$N(t) = \sum_{k=1}^K a_k e^{-b_k t} \ .$$

In applications, it must be $a_k > 0$ and $b_k \geq 0$. The special case in which Ω is a segment, $\Omega = (0, 1)$ has been studied in [94], using an interesting idea which we describe when $K = 2$. We introduce the auxiliary functions

$$r_1(t) = \int_0^t e^{-b_1(t-s)} w_x(s) ds \ , \qquad r_2(t) = \int_0^t e^{-b_2(t-s)} w_x(s) ds$$

and we note that

$$\frac{\partial}{\partial x} r_i = \int_0^t e^{-b_i(t-s)} w_{xx}(s) ds \ , \qquad r_i' = w_x - b_i r_i \ .$$

Hence, the equation

$$w' = \int\limits_0^t \left(a_1 e^{-b_1(t-s)} + a_2 e^{b_2(t-s)} \right) w_{xx}(s) ds$$

is equivalent to the system

$$w' = a_1 \frac{\partial}{\partial x} r_1 + a_2 \frac{\partial}{\partial x} r_2 , \quad r_1' = w_x - b_1 r_1 , \quad r_2' = w_x - b_2 r_2 .$$

We do not insist on this approach (used to study stability and spectrum in [38, 94]) which has not yet been exploited in the study of control problems (controllability when $N(t) = 1 + ae^{bt}$ is studied in [68] using moment methods, as in Chap. 5).

2.6 The Derivation of the Models

Among the several applications of systems with persistent memory, we keep in mind applications to thermodynamics and (nonfickian) diffusion, and to viscoelasticity. We give a short account of the derivation of the models (2.1) and (2.3) in these cases. See [20] for a detailed analysis.

2.6.1 Thermodynamics with Memory and Nonfickian Diffusion

We first recall the derivation of the memoryless heat equation in a bar, which follows from two fundamental physical facts: conservation of energy and the fact that the temperature is a measure of energy, i.e.,

$$e'(x, t) = -q_x(x, t) , \qquad \theta'(x, t) = \gamma e'(x, t) , \quad \gamma > 0 .$$

Here, q is the flux of heat, e is the density of energy and θ is the temperature.[3] So, we have

$$\theta' = -\gamma q_x(x, t) .$$

This equality is combined with a "constitutive law". Fourier law assumes that the flux responds immediately to changes in temperature:

$$q(x, t) = -k\theta_x(x, t) \tag{2.46}$$

---

[3] the minus sign because the total internal energy $\int_a^b e(x, t) dx$ of a segment (a, b) decreases when the flux of heat is directed to the exterior of the segment.

($k > 0$ and the minus sign because the flux is toward parts which have lower temperature). Combining these equalities we get the memoryless heat equation

$$\theta' = (k\gamma)\theta_{xx}.$$ (2.47)

The heat equation with memory is obtained when (as done in [39], following the special case examined in [13]) we take into account the fact that the transmission of heat is not immediate, and Fourier law is replaced with

$$q(x, t) = -k \int_{-\infty}^{t} N(t - s)\theta_x(x, s)ds,$$ (2.48)

which gives (the product $k\gamma N(t)$ is renamed $N(t)$)

$$\theta' = \int_{-\infty}^{t} N(t - s)\theta_{xx}(x, s)ds.$$ (2.49)

If the system is subject to the action of an external control f, for example if we impose

$$\theta(0, t) = f(t), \qquad \theta(\pi, t) = 0,$$

then the control acts after a certain initial time t_0 and we can assume $t_0 = 0$. So, we get the control problem

$$\theta' = \int_{0}^{t} N(t - s)\theta_{xx}(s)ds + H(t), \quad \theta(0, t) = f(t), \quad \theta(\pi, t) = 0.$$ (2.50)

The affine term $H(t) = H(x, t)$ takes into account the previous history of the system,

$$H(x, t) = \int_{-\infty}^{0} N(t - s)\theta_{xx}(x, s)ds$$

(we noted that in the study of control problems we can assume $H = 0$).

So, we get system (2.3).

A similar equation is obtained when θ represents the concentration of a solute in a solvent, and q represents the flux across the position x at time t. Then, the constitutive law (2.46) is the Fick law, which leads to (2.47) as the law for the variation in time and space of the concentration (denoted θ) but it is clear that the assumption that the flux of matter reacts immediately to the variation of concentration is even less acceptable and, in particular in the presence of complex molecular structures, the law (2.48) is preferred, leading to (2.49) as the equation for the concentration. In fact, in Sect. 2.6.3 we shall see a further reason for this choice.

2.6.2 Viscoelasticity

We consider the simple case of a finite string (of constant density $\rho > 0$), which in the undeformed configuration is on the segment $[a, b]$ of the horizontal axis. After a deformation (which we assume "small" in the sense specified below) the point in position $(x, 0)$ will be found in position $(x, w(x))$ (so, we assume negligible horizontal motion, a first "smallness" constraint).

We call $w(x, t)$ the "deformation" of the string at time t and position x.

A deformation by itself does not produce an elastic traction: for example, a translation $w(x) = h$, the same for every x, does not produce any traction in the string. An elastic traction appears when neighboring points undergo different deformations.

Let us fix a point x_0. A small segment $(x_0, x_0 + \delta)$ on the right of x_0 exerts a traction on $[a, x_0]$ if $w(x_0 + \delta) \neq w(x_0)$, i.e., $w(x_0 + \delta) - w(x_0) \neq 0$, and elasticity assumes that this traction is $k\,(w(x_0 + \delta) - w(x_0) + o\,(w(x_0 + \delta) - w(x_0)))$.

It is an experimental fact that $k = k(\delta)$ and, with an acceptable error,

$$k = \frac{k_0}{\delta}$$

(in order to produce the same deformation in a longer string a smaller force is required). The number k_0 is positive since if $\delta > 0$ and $w(x_0 + \delta) - w(x_0) > 0$, then the traction on $[a, x_0]$ exerted by the part of the segment $x > x_0$ points upwards.

We proceed assuming the conditions of linear elasticity, which are new "smallness" condition. The first one is that the effect of $o\,(w(x_0 + \delta) - w(x_0))$ is negligibly small, and we ignore it.

Now, we balance the momentum on a segment (x_0, x_1) and the exterior forces acting on this segment, which are its weight, $p = -\rho g(x_1 - x_0)$ (g is the acceleration of gravity) and the difference of the tractions in x_1 and in x_0. Hence, we have

$$\frac{\mathrm{d}}{\mathrm{d}t}\left(\rho \int_{x_0}^{x_1} w'(x, t)\mathrm{d}x\right) = -(x_1 - x_0)\rho g$$

$$+ \frac{k_0}{\delta}\left[(w(x_1 + \delta, t) - w(x_1, t)) - (w(x_0 + \delta, t) - w(x_0, t))\right],$$

We approximate the second line with $k_0\,(w_x(x_1, t) - w_x(x_0, t))$ (the last "smallness" condition):

$$\frac{d}{dt}\left(\rho \int_{x_0}^{x_1} w'(x, t) dx\right) = -(x_1 - x_0)\rho g + k_0 [w_x(x_1, t) - w_x(x_0, t)]. \quad (2.51)$$

Now, we divide both the sides with $x_1 - x_0$ and we pass to the limit for $x_1 \to x_0$. As x_0 is a generic abscissa of points of the string, we rename it x and we get the string equation

$$\rho w''(x, t) = -\rho g + k_0 w_{xx}(x, t). \quad (2.52)$$

A hidden assumption in this derivation is that the elastic traction appears at the same time as the deformations $[w(x_1 + \delta, t) - w(x_1, t)]$, $[w(x_0 + \delta, t) - w(x_0, t)]$, and that it adjusts itself instantaneously to the variation of the deformation. In a sense, this is a common experience: a rubber cord shortens abruptly when released. But, if the cord has been in the freezer for some time, it will shorten slowly: the effect of the traction fades slowly with time. This is taken into account as follows: at every time t the traction depends also on the configuration of the string on every previous time s. Hence, instead of $k_0 [w_x(x_1, t) - w_x(x_0, t)]$, the traction exerted on $[x_0, x_1]$ at time t *and due to the deformation at time s* is approximated with

$$\sigma(x_0, x_1, t, s) = k_0 [(w_x(x_1, s) - w_x(x_1, s - \delta)) - ((w_x(x_0, s) - w_x(x_0, s - \delta))]$$
$$= k_0 [w_x'(x_1, s) - w_x'(x_0, s)] \delta + o(\delta).$$

We approximate this expression with:

$$\sigma(x_0, x_1, t, s) = k_0(t, s) [w_x'(x_1, s) - w_x'(x_0, s)] \delta.$$

Most often, $k(t, s) = k(t - s)$ and $k(t)$ is a decreasing function (the effect of previous deformations fades with time).

All the tractions originating at every time $s < t$ "sum" to give the traction acting on the segment $[x_0, x_1]$ at time t. So, Eq. (2.51) is replaced with

$$\frac{d}{dt}\left(\rho \int_{x_0}^{x_1} w'(x, t) dx\right) = -(x_1 - x_0)\rho g$$

$$+ \int_{-\infty}^{t} k(t - s) [w_x'(x_1, s) - w_x'(x_0, s)] ds = -\rho g(x_1 - x_0)$$

$$+ k(0) [w_x(x_1, t) - w_x(x_0, t)] + \int_{-\infty}^{t} k'(t - s) [w_x(x_1, s)$$

$$- w_x(x_0, s)] d\tau.$$

We divide both the sides with $x_1 - x_0$ and we pass to the limit for $x_1 - x_0 \to 0$, as above. This gives Eq. (2.1).

Note that in this model the stress at position x is

$$\sigma(x, t) = k(0)w_x(x, s) + \int_{-\infty}^{t} k'(t - s)w_x(x, s)ds \;.$$

Remark 2.9 An implicit assumption in the previous derivation: the traction acting on the extremum x_1 of the segment $[x_0, x_1]$ is approximated with $k_0 w'_x(x_1, s)$: we disregard the traction, which may "diffuse" from distant points. And so, in Eq. (2.1) $w''(x, t)$ on the left side depends on the history of $w(x, s)$, at the *same* position x.

Finally, when studying control problems we can integrate on $(0, t)$, since we can assume that the system is at rest for $t < 0$.

2.6.3 The Special Case of the Telegraphers' Equation

The simplest example of a system with persistent memory is the system

$$u'(x, t) = \int_{0}^{t} e^{-\gamma(t-s)} u_{xx}(x, s)ds \qquad u(x, 0) = u_0(x)$$

(physical considerations imply $\gamma \geq 0$). We compute the derivative (in time) of both the sides. If $\gamma = 0$ we get the string equation (2.52) (with $\rho = k_0 = 0$ and without the affine term) while if $\gamma > 0$ we get the the TELEGRAPHER'S EQUATION

$$u''(x, t) = u_{xx}(x, t) - u'(x, t) \qquad u(x, 0) = u_0(x) \;, \quad u'(x, 0) = 0 \;.$$

So, it has an interest to contrast the properties of the wave equation, the telegrapher's equation and, in view of Sect. 2.6.1, also of the heat equation. The telegraphers' equation has been studied in details and, when $x \in \mathbb{R}$, formulas for the solutions are known, see [89, Ch. VII-2]. We are not going to discuss these formulas. We use them to represent the solutions in the case described below (Figs. 2.1 and 2.2 (right)).

We consider the case $x \in \mathbb{R}$ (no boundary, i.e., we assume that $x \in [a, b]$ and that $b - a$ is "very large" when compared with the interval of time during which the system is studied). We assign a discontinuous initial condition[4] to the wave, telegrapher's and heat equations, an we compare the corresponding solutions. The

---

[4] this is not realistic for an elastic or viscoelastic body. We should consider continuous initial conditions, which are not everywhere differentiable, but the qualitative facts described below are the same. For completeness, Fig. 2.2 (right) shows the case that the initial deformation is zero, with nonzero and discontinuous initial velocity.

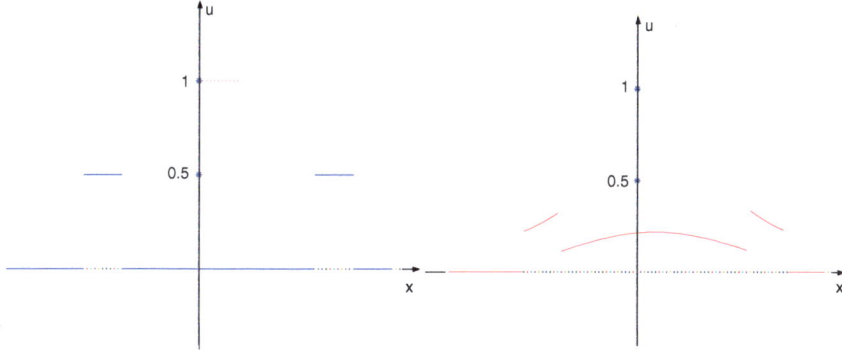

Fig. 2.1 *left* the graph of the initial condition (*dots*). When $t = 3$: the deformation computed from the (memoryless) wave equation (*left*) and from the telegrapher's equation (*right*)

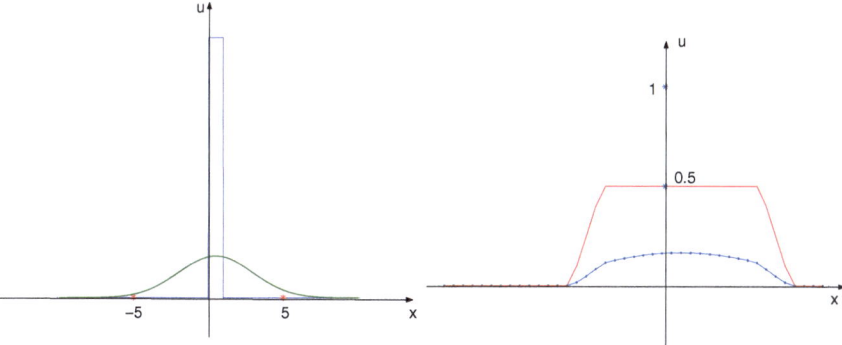

Fig. 2.2 Solution of the heat equation at the time $t = 3$ and the graph of the initial condition (*left*); wave and telegraph equations (at the time $t = 3$) with $u_0(x) = 0$ and $v_0(x) = 1$ for $x \in (0, 1)$ and $v_0(x) = 0$ otherwise (*right*)

equations are, respectively,

$$u'' = u_{xx}, \quad u'' = u_{xx} - u', \quad u' = u_{xx}$$

and the initial conditions are

$$u(x, 0) = u_0(x) = \begin{cases} 1 & \text{if } t \in (0, 1), \\ 0 & \text{otherwise} \end{cases} \qquad u'(x, 0) = v_0(x) = 0$$

(of course, in the case of the heat equation, we disregard the condition on $u'(x, 0)$).

We compare the graph of the function $x \mapsto u(x, 0) = u_0(x)$ and the graphs of the solutions at time $t = 3$, i.e., the maps $x \mapsto u(x, 3)$.

The noticeable facts for the *wave equation* are as follows:

- the solution at time $t = 3$ reached the intervals $[-3, -2]$ traveling to the left and $[3, 4]$ traveling to the right. During the times $t \in [0, 3)$ the solutions encountered every point in $(-2, 3)$, but the signal did not leave any memory of itself: once passed, its effect is abruptly forgotten.
- the discontinuity of the initial condition is preserved. There is a sharp *wavefront* both "in front" and "behind" the "traveling wave".

Instead, for the solution of the *telegraphers' equation:*

- the solution is a wave traveling with the same speed 1 as the solution of the wave equation: at time $T = 3$ it reaches the points $x = 4$ and $x = -3$. In particular, there is a discontinuity, i.e., a sharp forward "wavefront", at these points.
- the solution is discontinuous also at the points -2 and 3; i.e., it has the same jump points as the solution of the wave equation, but once the signal "encounters" a point x its memory *persists forever in that position* (in fact, its effect is attenuated with time, but never becomes identically zero).

We contrast also with the solution of the *heat equation,* whose key property is:

- the solution of the heat equation is of class C^∞ and *it is not zero for every x.* It soon becomes "very small" and it cannot be detected in the graph, but it is nonzero, since it is given by

$$\frac{1}{\sqrt{4\pi t}} \int\limits_0^1 e^{-(x-s)^2/4t} ds .$$

Consequence of this: signals travels with infinite speed in the case of the (memoryless) heat equation.

If we interpret w as the concentration of a "liquid" which diffuses in a polymer then we have:

- if the diffusion is according to Fick law, concentration varies smoothly and there is no visible separation between wet and dry regions.
- if the law of the diffusion is the telegraph equation then there is a sharp separation between wet and dry parts of the body, a fact experimentally verified in presence of complex molecular structure.

This last observation suggested to replace the heat equation with Eq. (2.3) to model diffusion in polymers, see [21].

2.7 Problems[5] to Chap. 2

2.1. Consider the following system of two ordinary differential equations:

$$u'' = -3u + w, \quad w'' = -3w + u .$$

Let the initial conditions be $u(0) = u_0$, $u'(0) = 0$, $w(0) = w_0$, $w'(0) = 0$. Represent the solution as

$$\begin{pmatrix} u(t) \\ w(t) \end{pmatrix} = R(t) \begin{pmatrix} u_0 \\ w_0 \end{pmatrix} .$$

Prove that $R(t)$ is a cosine operator in the sense that it verifies equality (2.9).

2.2. Show that the family of the operators $R_+(t)$ defined by

$$(R_+(t)\phi)(x) = \frac{1}{2} (\phi(x + t) + \phi(x - t)) \tag{2.53}$$

is a cosine operator in $L^2(\mathbb{R})$, in the sense that equality (2.9) holds for this family of operators. Decide whether it is possible to represent this cosine operator with a Fourier type expansion, as in (2.10).

2.3. Show that the series representation (2.10) for the cosine operator can be obtained by separation of variables. Write it explicitly in the case $\Omega = (0, \pi)$.

2.4. Write the series (2.10), which represents the cosine operator of the problem

$$u'' = u_{xx} + u_{yy} \quad (x, y) \in Q = (0, \pi) \times (0, \pi), \quad u = 0 \text{ on } \partial Q$$

(∂Q is not of class C^2, but the results we have seen extend to this case).

2.5. Prove the integration by parts formulas (2.11) and (2.12).

2.6. Prove Theorem 2.2 when $f = 0$, $G = 0$, but the initial conditions are not zero; and do the same when control and initial conditions are zero, but $G \neq 0$.

2.7. Prove that under the condition of Theorem 2.2, and if furthermore u_0 and v_0 belong to $\mathscr{D}(\Omega)$ then the function $u(t)$ is of class $C^\infty([0, T]; L^2(\Omega))$.

2.8. Let $x \in (0, \pi)$ and $u = u(x, t)$ while $w = w(t)$ depends only on the time. Assuming zero initial conditions, discuss the time at which the effect of the input f affects the output y in the case of the systems

$$
\begin{cases}
u'' = u_{xx}, \quad u(0, t) = f(t), \quad u(\pi, t) = 0 \\
w'' = -w + \int_{1/3}^{1/2} u(x, s)dx \\
y(t) = w(t),
\end{cases}
\qquad
\begin{cases}
w'' = -w + f(t) \\
u_{tt} = u_{xx} + \mathbf{1}_{(1/3, 1/2)}(x)w(t) \\
u(0, t) = u(\pi, t) = 0 \\
y(t) = \int_{3/4}^{1} u(x, t)dt
\end{cases}
$$

---

[5] Solutions at the address http://calvino.polito.it/~lucipan/materiale_html/P-S-CH-2.

The function $\mathbf{1}_{(1/3,1/2)}(x)$ is the characteristic function of $(1/3, 1/2)$.

2.9. The notations u and w are as in Problem 2.8 while $x \in (0, 7)$. Let $y(t) = w(t) \in \mathbb{R}$ be the output of the following system ($\alpha > 1$ is a real parameter):

$$u'' = u_{xx} + \mathbf{1}_{(1,\alpha)}(x) f(t) , \quad w'' = -w + \int_3^6 u(x, t) dx$$

($u = u(x, t)$ with $x \in (0, 7)$) and conditions

$$\begin{cases} u(0, t) = 0, \ u(7, t) = f(t) , \\ u(x, 0) = 0, \ u'(x, 0) = 0, \ w(0) = 0. \end{cases}$$

Study at what time the effect of the external input $f(t)$ will influence the observation $y(t)$ and specify whether the time depends on α.

2.10. On a region Ω, consider the problem

$$w' = \Delta w + \int_0^t \Delta w(s) ds, \quad w(0) = u_0 \in L^2(\Omega), \quad w = 0 \text{ on } \partial\Omega. \qquad (2.54)$$

This equation is not of the same type as those studied in this book and has different control properties, see [37, 41–43]. Prove that Eq. (2.54) can be reduced to a Volterra integral equation using the semigroup e^{At}.

2.11. Let $\Omega = (0, 1)$ and consider the heat equation with memory (2.50) with $H = 0$ and $N(t) \equiv 1$ (hence, an integrated version of the string equation). Assume zero initial condition and $\theta(0, t) = f(t)$, $\theta(1, t) = 0$. Use (2.48) with initial time $t_0 = 0$ and $k = 1$. Compute the flux $q(t)$ on $t \in (0, 2)$ and show that the function $t \mapsto q(t)$ belongs to $C([0, 2]; L^2(0, 1))$.

Let either $T = 1$ or $T = 2$. Study whether the pair $(\theta(T), q(T))$ can be controlled to hit any target $(\xi, \eta) \in L^2(0, 1) \times L^2(0, 1)$ using a control $f \in L^2(0, T)$.

Chapter 3
Moment Problems and Exact Controllability

3.1 The Goal of This Chapter and the Moment Problem

Controllability of linear systems can often be reduced to the solution of suitable, equivalent, moment problems. The moment problem in Hilbert spaces is studied in this chapter. This problem has many facets and we confine ourselves to those properties which are needed in the study of exact controllability for wave-like equations, stressing the relation of the moment operator with Riesz sequences. We present several examples both to clarify the properties of the moment operator and of the Riesz sequences and to show the applications to exact controllability. In particular, the last worked example presents in a simplified setting the crucial ideas used in the study of controllability of systems with persistent memory.

To introduce the moment problem, we fix a denumerable index set $\mathbb{J}$ and a sequence space Y. The sequences of Y are indexed by $\mathbb{J}$. When needed for clarity we write $Y(\mathbb{J})$. An abstract version of the moment problem is as follows: H is a separable[1] Hilbert space (inner product is $\langle \cdot, \cdot \rangle$ and the norm is $| \cdot |$. We write $\langle \cdot, \cdot \rangle_H$ and $| \cdot |_H$ when needed for clarity). The spaces Y and H are either real or complex linear spaces.

Let $\{e_n\}_{n\in\mathbb{J}}$ be a sequence in H. The MOMENT PROBLEM *respect to the index set* $\mathbb{J}$ *and the sequence* $\{e_n\}_{n\in\mathbb{J}}$ consists in the study of the map

$$\mathbb{M} : \quad f \mapsto \mathbb{M}f = \{\langle f, e_n \rangle_H\} = \{\langle f, e_n \rangle\} \tag{3.1}$$

and the problem of interest is to understand whether the map is surjective: given $\{c_n\} \in Y$, we would like to understand whether there exists at least one solution f of the sequence of the equations

$$\langle f, e_n \rangle = c_n, \quad n \in \mathbb{J} \tag{3.2}$$

---

[1] i.e., it contains a denumerable orthonormal basis.

© The Author(s) 2014
L. Pandolfi, *Distributed Systems with Persistent Memory*, SpringerBriefs in Control,
Automation and Robotics, DOI 10.1007/978-3-319-12247-2_3

and we would like a formula for a solution.

Note that in general the sequence $\{\langle f, e_n \rangle\}$ does not belong to Y and so

$$\text{dom } \mathbb{M} = \{f \in H : \mathbb{M}f \in Y\}. \tag{3.3}$$

For definiteness, unless clear from the contest, in this chapter the Hilbert spaces are real and $\mathbb{J} = \mathbb{N}$, but Problems 3.1, 3.2, 3.3 show that the properties of the moment problem depend[2] on the index set $\mathbb{J}$ which, in practice, is given by the problem under study.

In this chapter, we present only the properties which are needed to solve the control problems that we are going to study. Interested readers can look at the books [2, 15, 34, 45, 54, 57] where also the case that H is a Banach space is considered. In this case, the crochet denotes the duality pairing. Several moment problems, which are practically and historically important are formulated in this more general settings, see [56, 58, 87]. But, in the cases of our interest H will be a Hilbert space and

$$Y = l^2.$$

Then, we can state a first property of $\mathbb{M}$:

Lemma 3.1 *If the sequence $\{e_n\}$ is not finite, $Y = l^2$ and $\text{dom } \mathbb{M} = H$ then*

$$w - \lim e_n = 0. \tag{3.4}$$

Hence, $\{e_n\}$ is bounded.

In fact, for every $f \in H$ we have $\{\langle f, e_n \rangle\} \in l^2$ and so $\lim \langle f, e_n \rangle = 0$. So, (3.4) holds and we know that every weakly convergent sequence is bounded.

3.1.1 A Worked Example

In order to see the relation of controllability and moment problems, we discuss the simple example of the string equation on a *bounded interval*. For simplicity the interval is $(0, \pi)$. The string is controlled at one end by a square integrable control:

$$u'' = u_{xx}, \qquad u(0, t) = f(t), \qquad u(\pi, t) = 0 \tag{3.5}$$

(null initial condition $u(x, 0) = 0$ and $u'(x, 0) = 0$).

The operator A (Sect. 1.3) is now $Au = u_{xx}$ with $\text{dom } A = H^2(0, \pi) \cap H_0^1(0, \pi)$ and the elements of a complete orthonormal sequence of eigenvectors are

---

[2] More formally, we should say that the properties of the moment operator depend on the *sequence* $\{e_n\}$ and not solely on the set of the elements e_n.

$$\phi_n(x) = \sqrt{\frac{2}{\pi}} \sin nx .$$

Proceeding by separation of variables we see that

$$u(x,t) = \sum_{n=1}^{+\infty} u_n(t)\phi_n(x), \quad u_n(t) = \sqrt{\frac{2}{\pi}} \int_0^\pi u(x,t) \sin nx dx .$$

The function $u_n(t)$ solves

$$u_n''(t) = -n^2 u_n(t) + n\sqrt{\frac{2}{\pi}} f(t) .$$

So, we have

$$u_n(t) = \sqrt{\frac{2}{\pi}} \int_0^t \sin n(t-s) f(s) ds \quad \text{and} \quad u_n'(t) = n\sqrt{\frac{2}{\pi}} \int_0^t \cos n(t-s) f(s) ds .$$

Let us fix a target $\sum_{n=1}^{+\infty} \xi_n \phi_n(x)$. This target is reachable at a certain time T if there exists a solution $f \in L^2(0, T)$ of the moment problem

$$\int_0^T f(T-s) \sin ns ds = \sqrt{\frac{\pi}{2}} \xi_n . \tag{3.6}$$

The theory of Fourier sine series shows that solvability with square integrable controls requires $\{\xi_n\} \in l^2$, i.e., that controllability has to be studied in $L^2(0, \pi)$.

Let us represent also $u'(x, T) = \sum_{n=1}^{+\infty} \eta_n \phi_n(x)$. If we want to control both $u(x, T)$ and $u'(x, T)$ then we must find $f \in L^2(0, T)$, which solves the moment problem

$$\int_0^T f(T-s) (\sin ns) ds = \sqrt{\frac{\pi}{2}} \xi_n , \quad \int_0^T f(T-s) (\cos ns) ds = \sqrt{\frac{\pi}{2}} \frac{1}{n} \eta_n . \tag{3.7}$$

The theory of the Fourier series imposes both $\{\xi_n\} \in l^2$ and $\eta_n/n = c_n$ with $\{c_n\} \in l^2$. Hence, the targets for the velocity have the representation $\sum_{n=1}^{+\infty} c_n (n\phi_n(x))$ with $\{c_n\} \in l^2$. As proved in Sect. 1.3, we then have $\eta \in H^{-1}(0, \pi)$.

We examine solvability of these moment problems. We consider first controllability of the sole deformation $u(x, T)$, and then of the pair $(u(x, T), u'(x, T))$.

The theory of the *Fourier sine* series shows that:

- if $T < \pi$, then the moment operator in (3.6) is not surjective, and the system is not controllable. In fact, the reachable targets ξ are not even dense in $L^2(0, \pi)$;
- if $T = \pi$: the moment problem (3.6) is uniquely solvable and we have $u(x, \pi) = \xi(x)$ if

$$f(t) = \sqrt{\frac{2}{\pi}} \sum_{n=1}^{+\infty} \xi_n \sin n(\pi - t) = \xi(\pi - t). \tag{3.8}$$

So, when $T = \pi$, the transformation $f \mapsto \xi$ is bounded and boundedly invertible between $L^2(0, \pi)$, the space of the controls, and $L^2(0, \pi)$, the space of the targets ξ. Hence, any $\xi \in L^2(0, \pi)$ is reachable using a *unique* control $f \in L^2(0, \pi)$, which *depends continuously* on $\xi \in L^2(0, \pi)$ (note that this follows also from $u(x, \pi) = f(\pi - x)$, a fact easily seen directly from Eq. (3.5). Compare with Chap. 1).

- If $T > \pi$ the moment problem (3.6), hence the control problem is solvable and every target ξ can be reached using infinitely many different controls. The one of minimal norm is (the function $f(t)$ is that in (3.8)):

$$f_T(t) = \begin{cases} 0 & \text{if } 0 < t < T - \pi \\ f(t - (T - \pi)) & \text{if } T - \pi < t < T. \end{cases}$$

Hence, $f_T \in L^2(0, T)$ *depends continuously on* $\xi \in L^2(0, \pi)$.

If we want to control both the components u and u', then we must find $g(t) = f(T - t)$, which solves every equation in (3.7). We have:

- The theory of the Fourier series shows that this is possible only if $T \geq 2\pi$ and that for $T = 2\pi$ we can choose

$$f(2\pi - t) = g(t) = \frac{1}{\sqrt{2\pi}} \sum_{n=1}^{+\infty} (c_n \cos nt + \xi_n \sin nt), \quad c_n = \frac{1}{n} \eta_n.$$

This control is not unique: the function $f(t)$ given by $f(T - t) = g(t) + a$ solves the same moment problem as $f(t)$. When $a = 0$ we get the *control of minimal norm*, which steers the null initial condition to the target (ξ, η). *The control of minimal norm depends continuously on* $(\xi, \eta) \in L^2(0, \pi) \times H^{-1}(0, \pi)$.
- Note that the control $f(t) \equiv 1$ for $t \in [0, 2\pi]$ controls $(u(x, t), u'(x, t))$ from $(u(x, 0), u'(x, 0)) = (0, 0)$ to the same final position $(u(x, 2\pi), u'(x, 2\pi)) = (0, 0)$ in spite of the fact that at intermediate times it will be $(u(x, t), u'(x, t)) \neq (0, 0)$. For example, when $t = \pi$ we have $u(x, \pi) = f(\pi - x) = 1$.

In conclusion:

- these example justify the choice $Y = l^2$.

- controllability of the string equation is equivalent to the solution of certain moment problems. The known theory of the Fourier series completely clarifies solvability of the moment problems involved, hence controllability of the string equation;
- if the goal is controllability of the sole deformation $u(x, T)$ at time $T = \pi$, we can deduce controllability from the formula $u(x, \pi) = f(\pi - x)$, similar to what we did in Chap. 1, but here we have an important difference: in the case of the unbounded string we can control $u(x, T)$ (for $0 < x < T$), but not the pair $(u(x, T), u'(x, T))$, whatever the time T, see Remark 1.2. Controllability of the pair is possible in a bounded string (for T large enough) thanks to the *reflections* of the signal at the fixed, uncontrolled, end.

3.2 Properties of the Moment Operator When $Y = l^2$

We denote $\{\tilde{e}_k\}$ the standard basis of l^2, i.e., for every fixed k the element $\tilde{e}_k$ is the sequence $\{\delta_k^n\}_{n \in \mathbb{N}}$ where δ_k^n is the KRONECKER DELTA defined by

$$\delta_k^n = \begin{cases} 1 \text{ if } k = n \\ 0 \text{ if } k \neq n \end{cases} \quad \text{so that} \quad \langle \tilde{e}_n, \tilde{e}_r \rangle_{l^2} = \delta_n^r.$$

In our applications, the moment operator $\mathbb{M}$ defined in (3.1)–(3.3) will be continuous but for clarity we don't assume continuity from the outset.

If dom $\mathbb{M}$ is dense, we can compute $\mathbb{M}^*$ and we find

$$\mathbb{M}^*\{\alpha_k\} = \sum_{k=1}^{+\infty} \alpha_k e_k \quad \text{so that} \quad \mathbb{M}^* \tilde{e}_n = e_n. \tag{3.9}$$

The domain of $\mathbb{M}^*$ is the set of those sequences $\{\alpha_k\}$ for which the series in (3.9) converges weakly in H.

The formula for $\mathbb{M}^*$ is easily seen from the following equalities

$$\langle \mathbb{M}f, \{\alpha_k\} \rangle_{l^2} = \sum_{k=1}^{+\infty} \langle f, e_k \rangle_H \bar{\alpha}_k = \langle f, \sum_{k=1}^{+\infty} \alpha_k e_k \rangle_H.$$

Example 3.1 We see that the series of $\mathbb{M}^*\{\alpha_k\}$ in general is not norm convergent. Let us consider an orthonormal basis $\{x_n\}$ of H and let

$$e_1 = x_1, \quad e_n = n(x_n - x_{n-1}) \text{ if } n \geq 2.$$

We have, for every $f \in H$,

$$f = \sum_{n=1}^{+\infty} f_n x_n , \qquad \mathbb{M} f = \{f_1 , \, 2 \, (f_2 - f_1) , \, 3 \, (f_3 - f_2) , \ldots\} .$$

So, dom $\mathbb{M}$ is dense (since it contains every linear combination of the x_n) and $\mathbb{M}^*$ can be defined. The special sequence $\{1/k\}$ belongs to dom $\mathbb{M}^*$ since

$$\sum_{n=1}^{k} \alpha_k e_k = x_1 + \frac{1}{2} 2 \, (x_2 - x_1) + \frac{1}{3} 3 \, (x_3 - x_2) + \cdots \frac{1}{k} k \, (x_k - x_{k-1}) = x_k \rightharpoonup 0 .$$

But, $\{x_k\}$ is not convergent in the norm topology.

Theorem 3.1 *The operator $\mathbb{M}$ is closed.*

Proof We must prove that any sequence $\{(f_k, \mathbb{M} f_k)\} \in \mathscr{G}(\mathbb{M})$ (the graph of $\mathbb{M}$), if convergent, converges to a point of the graph. So, let

$$(f_k, \mathbb{M} f_k) \to (f_0, y_0) .$$

For every k, $\mathbb{M} f_k$ is a sequence, which we denote $\{\xi_k^n\}_{n \in \mathbb{N}}$ and y_0 is a sequence, $y_0 = \{\xi_0^n\}_{n \in \mathbb{N}}$. Our assumption is that

$$\lim_k f_k = f_0 \text{ in } H , \qquad \xi_k^n = \langle f_k, e_n \rangle_H , \qquad \lim_k \{\xi_k^n\} = \{\xi_0^n\} \text{ in } l^2$$

so that we have also

$$\lim_k \xi_k^n = \xi_0^n \quad \text{for all } n.$$

Hence, using the continuity of the inner product, for *every fixed n* we have

$$\langle f_0, e_n \rangle = \lim_k \langle f_k, e_n \rangle = \lim_k \xi_k^n = \xi_0^n .$$

Our assumption is that $\lim_k \{\xi_k^n\} = \{\xi_0^n\} = y_0$ in l^2. So:

$$y_0 = \{\xi_0^n\} = \lim_k \{\xi_k^n\} = \{\langle f_0, e_n \rangle\} \quad \text{i.e., } f_0 \in \text{dom } \mathbb{M} \text{ and } \mathbb{M} f_0 = y_0.$$

In conclusion, the sequence $(f_k, \mathbb{M} f_k) \in \mathscr{G}(\mathbb{M})$ converges to the point $(f_0, y_0) = (f_0, \mathbb{M} f_0)$ of the graph, as wanted.

So, we have the following important consequence:

Corollary 3.1 (1) *If $\mathbb{M}$ has closed domain, then it is continuous.* (2) *If $\mathbb{M}$ has closed image, in particular if it is surjective, and if it is invertible then $\mathbb{M}^{-1}$ is continuous.*

Every continuous operator can be extended by continuity to the closure of its domain. This extension is not really needed for the moment operator. In fact:

Lemma 3.2 *If $\mathbb{M}$ is continuous then its domain is closed.*

Proof Let $f_n \in \text{dom } \mathbb{M}$ and let $f_n \to f_0$. We prove $f_0 \in \text{dom } \mathbb{M}$, i.e., $\{\langle f_0, e_n \rangle\} \in l^2$. Continuity of $\mathbb{M}$ implies the existence of $\lim_n \mathbb{M} f_n$ in l^2:

$$\lim_n \mathbb{M} f_n = \lim_n \{\langle f_n, e_k \rangle\}_{k \in \mathbb{N}} = \{y_k\}_{k \in \mathbb{N}} \quad \text{convergence in } l^2.$$

Continuity of the inner product shows that

$$\langle f_0, e_k \rangle = \lim_n \langle f_n, e_k \rangle = y_k \quad \text{i.e., } \mathbb{M} f_0 = \{y_k\} \in l^2, \text{ hence } f_0 \in \text{dom } \mathbb{M}.$$

We say that two sequences in H, $\{e_n\}$ and $\{e'_n\}$, are BIORTHOGONAL to each other when

$$\langle e_n, e'_k \rangle = \delta_n^k = \begin{cases} 1 \text{ if } n = k \\ 0 \text{ otherwise.} \end{cases}$$

Every orthonormal sequence is biorthogonal to itself (in particular, the standard basis of l^2 is biorthogonal to itself). In general, a sequence does not have any biorthogonal sequence, or may have infinitely many. It is also possible that it has only one biorthogonal sequence. The following Theorem 3.3 clarifies these points. We need the following observation:

Remark 3.1 We have $\mathbb{M} f = 0$ if and only if $f \perp \text{cl span}\{e_n\}$, and so $\ker \mathbb{M} \neq 0$ if and only if $\{e_n\}$ is not complete in H. Replacing H with $H_0 = \text{cl span}\{e_n\}$, we can study the moment operator $\mathbb{M}_0$, which is the restriction of $\mathbb{M}$ to H_0. The properties of this operator, inherited by those of $\mathbb{M}$, are:

- the operators $\mathbb{M}$ and $\mathbb{M}_0$ have the same image and $\ker \mathbb{M}_0 = 0$;
- the operator $\mathbb{M}_0$ is closed.

Problem 3.7 shows that $\mathbb{M}_0$ might not be densely defined. This we assume explicitly:

> Assumption: the operator $\mathbb{M}_0$ is densely defined in H_0, hence it has an adjoint operator from l^2 to H_0, $\mathbb{M}_0^* \{\alpha_n\} = \sum_{n=1}^{+\infty} \alpha_n e_n \in H_0$.

The next results are consequences of Theorem 1.2. In fact, Theorem 3.2 is just a reformulation of Theorem 1.2 while Theorem 3.3 needs to be proved.

Theorem 3.2 *We have:*

1. *The moment problem (3.2) admits a unique solution $f \in H_0$ which depends continuously on $\{c_n\} \in \text{im } \mathbb{M} \subseteq l^2$ if and only if $\mathbb{M}_0^{-1}$ is continuous, which is the case if and only if there exists $M > 0$ such that*

$$\frac{1}{M}|f|_H^2 \le |\mathbb{M}_0 f|_{l^2}^2 = \sum_{n=1}^{+\infty} |\langle f, e_n \rangle|^2, \quad \forall f \in H_0. \tag{3.10}$$

2. *The moment problem(3.2) admits a solution for every $\{c_n\} \in l^2$ if and only if $\mathbb{M}_0$ is surjective, and this is the case if and only if there exists $m_0 > 0$ such that*

$$m_0|\{\alpha_n\}|_{l^2}^2 \le |\mathbb{M}_0^*\{\alpha_n\}|_H^2 \quad \text{i.e.,} \quad m_0 \sum_{n=1}^{+\infty} |\alpha_n|^2 \le \left|\sum_{n=1}^{+\infty} \alpha_n e_n\right|_H^2 \quad \forall \{\alpha_n\} \in l^2.$$

$$\tag{3.11}$$

3. *The moment problem admits a unique solution $f \in H_0$ for every $\{c_n\} \in l^2$ if and only if both the conditions(3.10) and(3.11) hold. This solution depends continuously on $\{c_n\}$.*

Now we prove:

Theorem 3.3 *Let the sequence $\{e_n\}$ admit at least one biorthogonal sequence $\{\xi_n\}$. Then, we have:*

1. *for every k, we have $\tilde{e}_k \in \text{im } \mathbb{M}$ ($\{\tilde{e}_k\}$ is the standard basis of l^2).*
2. *the sequence $\{e_n\}$ admits one and only one biorthogonal sequence $\{e'_n\}$ whose elements belong to $H_0 = \text{cl span}\{e_n\}$.*
3. *if the sequence $\{e_n\}$ is complete in H then the biorthogonal sequence is unique.*
4. *we have $|e'_n|_H \le |\xi_n|_H$ for every n.*
5. *if $\mathbb{M}_0$ has dense image ($\mathbb{M}_0$ is the operator in Remark 3.1) then the biorthogonal sequence $\{e'_n\}$ is given by*

$$e'_n = (\mathbb{M}_0)^{-1} \tilde{e}_n. \tag{3.12}$$

6. *if $\text{im } \mathbb{M} = l^2$ then the sequence $\{e_n\}$ admits a biorthogonal sequence which is bounded.*

Proof Property 1 holds since $\tilde{e}_k = \{\delta_n^k\} = \{\langle \xi_n, e_k \rangle\} = \mathbb{M}\xi_n$. We prove 2. Existence is clear: let P be the orthogonal projection of H onto H_0. Then,

$$e'_n = P\xi_n$$

is a sequence in H_0 and

$$\langle e_k, e'_n \rangle = \langle e_k, \xi_n \rangle = \delta_k^n.$$

We prove unicity. Let $\{e''_n\}$ be a second biorthogonal sequence which belongs to H_0. Then, $\{e'_n - e''_n\}$ belongs to H_0. Furthermore, for every k, the case $k = n$ included, we have

$$\langle e'_n - e''_n, e_k \rangle = 0$$

so that *for every n we have* $e'_n - e''_n \in H_0 \cap H_0^{\perp}$, i.e., $e'_n = e''_n$. If in particular $H_0 = H$, then the biorthogonal sequence in H is unique, i.e., property 3 holds.

We prove 4. Unicity of $\{e'_n\}$ in particular implies $P\xi_n = e'_n$ so that $|e'_n| = |P\xi_n| \leq |\xi_n|$.

We prove 5. The operator $\mathbb{M}_0 : H_0 = \mathrm{cl\,span}\{e_n\}$ to l^2 is invertible, the image is dense and by assumption its domain is dense. So, $\mathbb{M}_0^*$ and $(\mathbb{M}_0^*)^{-1}$ both exist. Formula (3.9) applied to $\mathbb{M}_0$ gives $\tilde{e}_n = (\mathbb{M}_0^*)^{-1} e_n$. Let $e'_n = \mathbb{M}_0^{-1} \tilde{e}_n \in H_0$. It is easily seen that $\langle e_n, e'_k \rangle = \delta_n^k$. In fact:

$$\langle e_n, e'_k \rangle_H = \langle e_n, \mathbb{M}_0^{-1} \tilde{e}_k \rangle_H = \left\langle \left(\mathbb{M}_0^{-1}\right)^* e_n, \tilde{e}_k \right\rangle_{l^2} = \langle \tilde{e}_n, \tilde{e}_k \rangle_{l^2} = \delta_n^k.$$

Finally, we prove 6. If $\mathbb{M}$ is surjective, then also $\mathbb{M}_0$ is surjective. Hence, $\mathbb{M}_0^{-1}$ is continuous, see the second statement in Corollary 3.1. The continuity of $\mathbb{M}_0^{-1}$ implies that $\{e'_n\} = \left\{\mathbb{M}_0^{-1} \tilde{e}_n\right\}$, is bounded.

3.3 Riesz Sequences and Moment Problems

The strongest and most favorable case for a moment problem is when the restriction $\mathbb{M}_0$ of $\mathbb{M}$ to $H_0 = \mathrm{cl\,span}\{e_n\}$ is defined on H_0 (so it is continuous) and surjective (hence $\mathbb{M}_0^{-1} \in \mathscr{L}(l^2, H_0)$) and so when both the conditions (3.10) and (3.11) hold.

Continuity of $\mathbb{M}_0$ implies continuity of $\mathbb{M}_0^*$; surjectivity implies $(\mathbb{M}_0^*)^{-1} \in \mathscr{L}(H, l^2)$. So, $e_n = \mathbb{M}_0^* \tilde{e}_n$ and $e'_n = \mathbb{M}_0^{-1} \tilde{e}_n$ (see formulas (3.9) and (3.12)) and a solution of the moment problem (3.2) is

$$f = \sum_{n=1}^{+\infty} c_n e'_n. \tag{3.13}$$

In fact, the series converges for every $\{c_n\} \in l^2$ since

$$\sum_{n=1}^{+\infty} c_n e'_n = \sum_{n=1}^{+\infty} c_n \mathbb{M}_0^{-1} \tilde{e}_n = \mathbb{M}_0^{-1} \sum_{n=1}^{+\infty} c_n \tilde{e}_n.$$

The equality $\langle f, e_n \rangle_H = c_n$ follows from the continuity of the inner product:

$$\left\langle \sum_{k=1}^{+\infty} c_k e'_k, e_n \right\rangle = \sum_{k=1}^{+\infty} c_k \langle e'_k, e_n \rangle = c_n.$$

In particular, *the solution $f \in H_0$ of the moment problem is unique and depends continuously on $\{c_n\} \in l^2$.*

The previous observations suggest the following definition:

Let $K \subseteq \mathbb{N}$ be a set of indices, and let $l^2 = l^2(K)$. A sequence $\{e_n\}_{n \in K}$ in the Hilbert space H is a RIESZ SEQUENCE when the operator $\mathbb{M}_0$ from cl span $\{e_n\}_{n \in K}$ to $l^2(K)$ has both the following properties: (1) it is defined on H_0 (and so it is continuous); (2) it is surjective (hence, it has a continuous inverse).

   If $\{e_n\}_{n \in K}$ is a Riesz sequence in H and $H_0 = H$ then $\{e_n\}_{n \in K}$ is a RIESZ BASIS of H.

Remark 3.2 We note:

- the previous definition applies in particular if the set K is finite, but the case that $\{e_n\}$ is the basis of a finite dimensional subspace is not of much use in control theory, and so we shall pursue our study in the case that $\{e_n\}$ *is an infinite sequence*.
- if K is infinite, we can always assume that $K = \mathbb{N}$ but in the proofs of Sect. 3.4 it is convenient to use sets $K \neq \mathbb{N}$. However, we need some care, as seen in Problem 3.8.

   We sum up:

Theorem 3.4 *The moment operator is defined and continuous on H_0, the moment problem(3.2) is solvable for every $\{c_n\} \in l^2$ and the solution $f = \mathbb{M}_0^{-1}\{c_n\}$ is a continuous function of $\{c_n\} \in l^2$ if and only if $\{e_n\}$ is a Riesz sequence. The solution $f = \mathbb{M}_0^{-1}\{c_n\}$ is explicitly given by(3.13).*

Example 3.2 The following examples show that the conditions in the definitions of Riesz sequences are independent (see also Problem 3.7).

1. It is possible that $H_0 = H$ and that $\mathbb{M}$ be bounded and invertible, with closed image, which is not all of l^2. This is the case of the sequence $\{0, e_1, e_2, \ldots\}$ where $\{e_n\}$ is an orthonormal basis of H.
2. there exist moment operators $\mathbb{M}$ with dense domain in H, surjective and invertible (hence with continuous inverse), but not continuous. Let $H = L^2(0, \pi)$ and let $e_n = n \sin nx$ so that $H_0 = L^2(0, \pi)$ and, from the theory of the sine Fourier series, ker $\mathbb{M} = 0$. The function $f(x) = \sum_{n=1}^{+\infty}(2c_n/(n\pi)) \sin nx$ solves $\mathbb{M}f = c_n$ for every $\{c_n\} \in l^2$ and so $\mathbb{M}$ is surjective. But, the operator $\mathbb{M}$ is unbounded on $L^2(0, \pi)$.
3. the operator $\mathbb{M}$ can be bounded, defined on H and invertible, but with unbounded inverse. The example is similar to the previous one, with $H = L^2(0, \pi)$ and $e_n = (1/n) \sin nx$.

We investigate further the properties of the Riesz sequences. First we prove:

Theorem 3.5 *The following properties hold:*

1. *any Riesz sequence converges weakly to zero, hence it is bounded.*
2. *the sequence $\{e_n\}$ is a Riesz sequence if and only if the operator $\mathbb{M}_0^*$ is continuous on l^2, and continuously invertible.*
3. *if $\{e_n\}$ is a Riesz sequence in H then $\sum_{n=1}^{+\infty} \alpha_n e_n$ converges in the norm of H if and only if $\{\alpha_n\} \in l^2$.*
4. *any Riesz sequence admits the biorthogonal sequence $\{e_n'\}$ in H_0, $e_n' = \left(\mathbb{M}_0^{-1}\right) \tilde{e}_n$.*
5. *if $\{e_n\}$ is a Riesz sequence, any $f \in H_0 = \mathrm{cl\ span}\{e_n\}$ has the* unique *expansion*

$$f = \sum_{n=1}^{+\infty} \alpha_n e_n, \qquad \{\alpha_n\} \in l^2. \tag{3.14}$$

6. *if $\{e_n\}$ is a Riesz sequence in H and $T \in \mathcal{L}(H, Y)$ has a bounded inverse, then $\{T e_n\}$ is a Riesz sequence in Y. In particular, $\{e_n\}$ is a Riesz sequence if and only if it is the image of an orthonormal basis $\{\chi_n\}$ of any Hilbert space, under a linear and continuous transformation, which has continuous inverse. So, if $T \in \mathcal{L}(H, Y)$ is surjective and boundedly invertible and if $\{e_n\}$ is a Riesz basis of H, then $\{T e_n\}$ is a Riesz basis of Y.*

Proof Let us restrict to H_0. If $\{e_n\}$ is Riesz then $\mathbb{M}_0$ is defined on H_0 and Lemma 3.1 implies that $\{e_n\}$ converges weakly to zero, hence it is bounded, in the Hilbert space H_0. These same properties hold in $H = H_0 \oplus H_0^\perp$.

We prove Property 2. If $\{e_n\}$ is Riesz then $\mathbb{M}_0$ is defined on H_0 and both $\mathbb{M}_0$ and $\mathbb{M}_0^{-1}$ are bounded. This happens if and only if $\mathbb{M}_0^*$ has these same properties (see Theorem 1.2).

Conversely, the existence of $\mathbb{M}_0^*$ implies that $\mathbb{M}_0$ has dense domain. Again using Theorem 1.2 we deduce that $\mathbb{M}_0$ and $\mathbb{M}_0^{-1}$ are bounded and Lemma 3.2 shows dom $\mathbb{M}_0 = H_0$. Hence, $\{e_n\}$ is a Riesz sequence.

We prove property 3. Using that $\mathbb{M}_0^*$ is continuous with continuous inverse, and $\left(\mathbb{M}_0^*\right)^{-1} e_n = \tilde{e}_n$, we see that the series $\sum_{n=1}^{+\infty} \alpha_n e_n$ converges if and only if $\left(\mathbb{M}_0^*\right)^{-1} \sum_{n=1}^{+\infty} \alpha_n e_n = \sum_{n=1}^{+\infty} \alpha_n \tilde{e}_n$ converges. This is the case if and only if $\{\alpha_n\} \in l^2$, since $\{\tilde{e}_n\}$ is an orthonormal basis.

Property 4 follows from property 5 in Theorem 3.3.

Now we prove property 5. The unicity of the expansion is clear since $\sum_{n=1}^{+\infty} \alpha_n e_n = 0$ implies $0 = \left(\mathbb{M}_0^*\right)^{-1} \sum_{n=1}^{+\infty} \alpha_n e_n = \sum_{n=1}^{+\infty} \alpha_n \tilde{e}_n$ and so $\{\alpha_n\} = 0$. Existence of the representation follows since $\mathbb{M}_0^*$ is surjective and so, from (3.9), any $f \in H_0$ has the representation

$$f = \mathbb{M}_0^*\{\alpha_n\} = \sum_{n=1}^{+\infty} \alpha_n e_n.$$

The first statement in Property 6 follows since the moment operator of the sequence $\{y_n\} = \{Te_n\}$ on cl span$\{y_n\}$ is $\mathbb{M}_0 T^*$. The second statement follows from the first one and (3.9). The last statement follows since the assumptions imply $TH = Y$.

Property 3 in Theorem 3.5 shows that in the study of Riesz sequences, we can confine ourselves to sequences which admit biorthogonal sequences. We consider the properties of the biorthogonal sequence $\{e_n'\}$ which belongs to H_0.

Theorem 3.6 *Let $\{e_n\}$ be a sequence in H_0 which admits a biorthogonal sequence $\{e_n'\}$ in H_0. The sequence $\{e_n\}$ is a Riesz sequence if and only if $\{e_n'\}$ is a Riesz sequence. Furthermore we have*

$$e_n' = \mathbb{M}_0^{-1} \tilde{e}_n = \mathbb{M}_0^{-1} \left(\mathbb{M}_0^*\right)^{-1} e_n \tag{3.15}$$

and every element $f \in H_0$ has the unique *expansion*

$$f = \sum_{n=1}^{+\infty} \tilde{\alpha}_n e_n', \quad \{\tilde{\alpha}_n\} \in l^2. \tag{3.16}$$

Proof We assume that $\{e_n\}$ is Riesz and we prove that $\{e_n'\}$ is Riesz. The converse implication is obtained interchanging the roles of the sequences.

From (3.12) we have $e_n' = (\mathbb{M}_0)^{-1} \tilde{e}_n$. The assumption that $\{e_n\}$ is a Riesz sequence shows that $\mathbb{M}_0$ is defined on H_0, it is continuous with continuous inverse. So, $\{e_n'\}$ is a Riesz sequence, thanks to property 6 in Theorem 3.5.

The representation (3.16) is (3.14) written in terms of $\{e_n'\}$.

The fact that $\mathbb{M}_0^{-1} \left(\mathbb{M}_0^*\right)^{-1}$ is bounded with bounded inverse gives:

Corollary 3.2 *If $\{e_n\}$ is a Riesz sequence in H_0 then the biorthogonal sequence $\{e_n'\}$ in H_0 has the property that cl span $\{e_n'\} = H_0$.*

A consequence is that we can equivalently characterize the property of being Riesz in terms of the sequence or of a suitable biorthogonal sequence; in terms of $\mathbb{M}_0$ or of the moment operator $\mathbb{M}_0'$ of $\{e_n'\}$:

Corollary 3.3 *Let $\{e_n\}$ be a sequence in H and let $H_0 = $ cl span $\{e_n\}$. The following properties are equivalent:*

1. *the sequence $\{e_n\}$ is a Riesz sequence;*
2. *the operator $\mathbb{M}_0$ defined on H_0 is continuous and continuously invertible i.e., there exist $A > 0$ and $B > 0$ such that*

$$A|f|^2 \le \sum_{n=1}^{+\infty} |\langle f, e_n \rangle|^2 \le B|f|^2 \quad \forall f \in H_0; \tag{3.17}$$

3. *the sequence $\{e_n\}$ admits a biorthogonal sequence $\{e_n'\}$ in H_0 which is Riesz;*

4. *the moment operator* $\mathbb{M}'_0$ *of* $\{e'_n\}$ *is defined on* H_0, *continuous and continuously invertible i.e., there exist* $A' > 0$ *and* $B' > 0$ *such that*

$$A'|f|^2 \leq \sum_{n=1}^{+\infty} |\langle f, e'_n \rangle|^2 \leq B'|f|^2 \quad \forall f \in H_0. \tag{3.18}$$

Finally, we characterize Riesz sequences without using the moment operator:

Theorem 3.7 *The following properties are equivalent:*

1. *the sequence* $\{e_n\}$ *is a Riesz sequence in* H;
2. *there exist numbers* $m_0 > 0$ *and* M *such that the following inequalities hold for every* $\{\alpha_n\} \in l^2$:

$$m_0 \left(\sum_{n=1}^{+\infty} |\alpha_n|^2 \right) \leq \left| \sum_{n=1}^{+\infty} \alpha_n e_n \right|_H^2 \leq M \left(\sum_{n=1}^{+\infty} |\alpha_n|^2 \right). \tag{3.19}$$

3. *inequalities (3.19) hold for any sequence* $\{\alpha_n\}$ *of scalars which has only* finitely *many nonzero elements.*

Proof Clearly, (3.19) holds for every $\{\alpha_n\} \in l^2$ if and only if it holds for those sequences which have only finitely many nonzero elements; i.e., if the series are finite sums. So, we prove that $\{e_n\}$ is a Riesz sequence if and only if (3.19) holds.

First, we assume that $\{e_n\}$ is a Riesz sequence and we prove that both the inequalities in (3.19) hold. By assumption, $\{e_n\}$ admits a biorthogonal sequence $\{e'_n\}$ in H_0, which is a Riesz sequence too. We represent any $f \in H_0$ as $f = \sum_{n=1}^{+\infty} \alpha_n e_n$. We use both the inequalities in (3.18) and we get:

$$\sum_{n=1}^{+\infty} |\alpha_n|^2 = \sum_{n=1}^{+\infty} |\langle f, e'_n \rangle|^2 \leq B'|f|^2 \leq \frac{B'}{A'} \sum_{n=1}^{+\infty} |\langle f, e'_n \rangle|^2 \leq \frac{B'}{A'} \sum_{n=1}^{+\infty} |\alpha_n|^2.$$

This is (3.19) with $m_0 = 1/B'$ and $M = 1/A'$ since $|f|^2 = \left| \sum_{n=1}^{+\infty} \alpha_n e_n \right|^2$.

Conversely, let the inequalities (3.19) hold. We define the operator K from l^2 to H_0 given by

$$K \sum_{n=1}^{+\infty} \alpha_n \tilde{e}_n = \sum_{n=1}^{+\infty} \alpha_n e_n \quad \text{so that } K \tilde{e}_n = e_n.$$

The property $e_n \in \text{im } K$ for every n implies that im K is dense in H_0.

The inequality on the right of (3.19) implies that K is continuous on l^2. The inequality on the left of (3.19) implies that K is invertible with continuous inverse. In particular its image is closed, hence equal to H_0. We sum up: $e_n = K \tilde{e}_n$ and K is bounded and boundedly invertible, hence $\{e_n\}$ is a Riesz sequence, thanks to item 6 of Theorem 3.5.

Corollary 3.4 *Any subsequence of a Riesz sequence is a Riesz sequence too.*

Proof In fact, the inequalities in (3.19) for the subsequence are the special case of (3.19) for the whole sequence: put $\alpha_n = 0$ when the index does not appear in the subsequence.

Most of the books on Riesz bases, which do not focus on moment problems, uses the inequalities in (3.19) as the definition of Riesz sequences.

We conclude with a property which is not elementary (see [45, Theorem 7.13]):

Theorem 3.8 *The bounded sequence $\{e_n\}$ is a Riesz sequence in H if and only if every $h \in H$ has a unique representation $h = \sum_{n=1}^{+\infty} c_n e_n$, and furthermore this series is unconditionally convergent.*

We recall that a series is UNCONDITIONALLY CONVERGENT when any series, which has the same elements in a different order is convergent to the same sum.

Remark 3.3 In the following study of exact controllability we shall use the properties of Riesz sequences. Different classes of sequences arise when studying different control problems, especially for equations of parabolic type, see [2].

3.4 Perturbations of Riesz Sequences

In applications, we need to prove that a given sequence is Riesz. The idea is to compare it with a "simpler" one, thanks to the fact that the property of being Riesz is preserved under "small" perturbations. We present two results along these lines, which combine the PALEY- WIENER and (part of) the BARI Theorems (the complete statement of Bari Theorem will not be used and is in Problem 3.15).

Note that it is not restrictive to use the inequalities (3.19) with $m_0 < 1$ and $M > 1$. This we do in the next proofs.

Theorem 3.9 *Let $\{e_n\}_{n\geq 1}$ be a Riesz sequence in a Hilbert space H and let $\{z_n\}_{n\geq 1}$ satisfy*

$$\sum_{n=1}^{+\infty} |e_n - z_n|^2 = \gamma < +\infty. \tag{3.20}$$

Then, we have:

1. *there exists $N \geq 1$ such that $\{z_n\}_{n\geq N}$ is a Riesz sequence;*
2. *If $\gamma < m_0$ where $m_0 < 1$ is the constant in (3.19), then $\{z_n\}_{n\geq 1}$ is a Riesz sequence;*
3. *the series $\sum_{n=1}^{+\infty} c_n z_n$ converges if and only if $\{c_n\} \in l^2$.*

Proof We fix any $\tilde{\gamma} < 1$. We choose any N such that

$$\sum_{n=N}^{+\infty} |e_n - z_n|^2 < \tilde{\gamma} m_0. \tag{3.21}$$

We drop the elements e_n and z_n with $n < N$. Let $\{x_n\}$ be an orthonormal basis of $\left[\text{span}\,\{e_n\}_{n \geq N}\right]^{\perp}$. We consider the sequence whose elements are the vectors x_n and e_n in any order, and the sequence whose elements are x_n and z_n in the same order. We denote $\{\hat{e}_n\}$ and $\{\hat{z}_n\}$ these (extended) sequences. The new sequence $\{\hat{e}_n\}$ is a Riesz basis of H and the inequalities (3.19) and (3.21) hold for the new extended sequences, with the same constants (here $m_0 < 1$, $M > 1$ is used).

We prove that $\{\hat{z}_n\}_{n \geq N}$ is a Riesz basis so that its subsequence whose elements are the vectors z_n with $n \geq N$ is a Riesz sequence, thanks to Corollary 3.4.

We define the following operator T: for every *finite* sequence $\{c_n\}$ we put

$$T\left(\sum c_n \hat{e}_n\right) = \sum c_n \hat{z}_n$$

so that, in particular, $T\hat{e}_n = \hat{z}_n$.

We prove that T is bounded, so that it admits a unique bounded extension to H. In fact,

$$\left|T\left(\sum c_n \hat{e}_n\right)\right| = \left|\sum c_n\left[(\hat{z}_n - \hat{e}_n) + \hat{e}_n\right]\right| \leq \left|\sum c_n(\hat{z}_n - \hat{e}_n)\right| + \left|\sum c_n \hat{e}_n\right|.$$

Schwarz inequality shows that

$$\left|\sum c_n(\hat{z}_n - \hat{e}_n)\right|^2 \leq \left(\sum |c_n|^2\right)\left(\sum |\hat{z}_n - \hat{e}_n|^2\right)$$

$$\leq \tilde{\gamma} m_0 \left(\sum |c_n|^2\right) \leq \tilde{\gamma} \left|\sum c_n \hat{e}_n\right|^2$$

(the last equality uses the left inequality in (3.19)). Hence, T is bounded, the norm is less then $\left(1 + \sqrt{\tilde{\gamma}}\right)$, and it can be extended[3] by continuity to H:

$$T\left(\sum_{n=1}^{+\infty} c_n \hat{e}_n\right) = \sum_{n=1}^{+\infty} c_n \hat{z}_n.$$

In order to prove that T is boundedly invertible we compare T with the identity transformation I and we use the fact that if $\|T - I\| < 1$ then T is surjective and boundedly invertible. We have

---

[3] Statement 5 of Theorem 3.5 shows that every element of H can be represented as $\sum_{n=1}^{+\infty} c_n \hat{e}_n$.

$$\left| (T - I) \left(\sum_{n=1}^{+\infty} c_n \hat{e}_n \right) \right|^2 = \left| \sum_{n=1}^{+\infty} c_n (\hat{z}_n - \hat{e}_n) \right|^2 \le \left(\sum_{n=1}^{+\infty} |c_n|^2 \right) \sum_{n=1}^{+\infty} |\hat{z}_n - \hat{e}_n|^2$$

$$= \left(\sum_{n=1}^{+\infty} |c_n|^2 \right) \sum_{n=N}^{+\infty} |z_n - e_n|^2 \le (\tilde{\gamma} m_0) \left(\sum_{n=1}^{+\infty} |c_n|^2 \right)$$

$$\le \tilde{\gamma} \left| \sum_{n=1}^{+\infty} c_n \hat{e}_n \right|^2 .$$

Hence, $\| T - I \| \le \tilde{\gamma} < 1$ and T is surjective and boundedly invertible. In conclusion, $\{\hat{z}_n\}_{n \ge N}$ is the image of the Riesz sequence $\{\hat{e}_n\}_{n \ge N}$ under a surjective and boundedly invertible transformation in H. Hence, it is a Riesz sequence too, see the statement 6 in Theorem 3.5. This proves property 1.

If $\gamma < m_0$ then in the previous proof we can take $N = 1$ and we get Property 2. Property 3 follows from item 3 in Theorem 3.5.

When condition (3.20) holds, we say that the two sequences $\{e_n\}$ and $\{z_n\}$ are QUADRATICALLY CLOSE.

Note that the property of being quadratically close is preserved if we change a finite number of elements of one or of both the sequences. Instead, the properties of being a Riesz sequence is not preserved (for example, it is not preserved if e_2 is replaced with $2e_1$).

Now, we add a condition under which $\{z_n\}_{n \ge 1}$, and not only the "tail" $\{z_n\}_{n \ge N}$, is a Riesz sequence. We define:

Definition 3.1 A sequence $\{z_n\}$ in a Hilbert space H is ω- INDEPENDENT when the following holds: if $\{\alpha_n\} \in l^2$ and if $\sum_{n=1}^{+\infty} \alpha_n z_n = 0$ (convergence in the norm of H) then we have $\alpha_n = 0$ for every n.

The left inequality in (3.19) shows that *any Riesz sequence is ω-independent*. We are going to prove that an ω-independence sequence which is quadratically close to a Riesz sequence is a Riesz sequence too (see Problem 3.15 to examine the case that $\{e_n\}$ is a Riesz *basis*).

Theorem 3.10 *Let $\{e_n\}$ be a Riesz sequence in the Hilbert space H and $\{z_n\}$ be quadratically close to $\{e_n\}$. If $\{z_n\}$ is ω-independent then it is a Riesz sequence.*

Proof In this proof, the l^2 space of the sequences $\{c_n\}_{n \ge N}$ is denoted $l^2([N, +\infty))$. We proved the existence of N such that $\{z_n\}_{n \ge N}$ is a Riesz sequence. Assuming that the sequence if ω-independent, we first prove that $\{z_n\}_{n \ge N-1}$ is a Riesz sequence too.

Let $H_N = \text{cl span} \{z_n\}_{n \ge N}$ and $H_{N-1} = \text{cl span} \{z_n\}_{n \ge N-1}$. Of course, $H_{N-1} \supseteq H_N$ and the assumption that $\{z_n\}$ is ω-independent implies $H_{N-1} \ne H_N$ and so there exists $\hat{g} \ne 0$ in H_{N-1} such that $\hat{g} \perp H_N$.

Note that

$$\langle \hat{g}, z_{N-1} \rangle \neq 0$$

because $\hat{g} \in H_{N-1}$ is not zero.

Let $\mathbb{M}_N \colon H_N \mapsto l^2([N, +\infty))$ be the moment operator of $\{z_n\}_{n \geq N}$ so that $\mathbb{M}_N$ is defined on H_N, continuous and with continuous inverse and

$$\langle f, z_n \rangle = c_n, \quad n \geq N \iff f = f_N + g, \quad g \in H_N^\perp, \quad f_N = \mathbb{M}_N^{-1}\{c_n\}.$$

Now, we prove that $\{z_n\}_{n \geq N-1}$ is a Riesz sequence. We consider the moment problem

$$\langle f, z_n \rangle = c_n, \quad n \geq N - 1$$

and we search for a solution in H_{N-1}, hence of the form $f_N + \alpha \hat{g}$. The solution exists, given by

$$f_N + \frac{c_{N-1} - \langle f_N, z_{N-1} \rangle}{\langle \hat{g}, z_{N-1} \rangle} \hat{g}.$$

It is clear that the solution depends continuously on $\{c_n\}_{n \geq N-1}$.

In conclusion, $\mathbb{M}_{N-1} \colon H_{N-1} \mapsto l^2([N-1, +\infty))$ is surjective (with continuous inverse). In order to finish the proof that $\{z_n\}_{n \geq N-1}$ is a Riesz sequence, we must prove that $\mathbb{M}_{N-1}$ is defined on H_{N-1}, i.e., that $\mathbb{M}_{N-1} f \in l^2([N-1, +\infty))$ for every $f \in H_{N-1}$. But this is clear since when $f \in H_{N-1}$ the elements of the sequence $\mathbb{M}_{N-1} f$ are $\langle f, z_{N-1} \rangle$ followed by the elements of the sequence $\mathbb{M}_N f$ which belongs to $l^2([N, +\infty))$ by assumption.

The argument can be iterated and after a finite number of steps we get that $\{z_n\}_{n \geq 1}$ is a Riesz sequence.

Finally, we prove:

Theorem 3.11 *Let $\{e_n\}_{n \in \mathbb{N}}$ be such that $\{e_n\}_{n \geq N}$ is a Riesz sequence. Then, im $\mathbb{M}$ has finite codimension and it is is closed.*

Proof Recall the notation $l^2(K)$ to denote the Hilbert space of the sequences for which $\sum_{n \in K} |\alpha_n|^2 < +\infty$ so that

$$l^2 = l^2([1, N-1]) \oplus l^2([N, +\infty)).$$

We can replace H_0 to H. So, it is not restrictive to assume $H_0 = \text{cl span } \{e_n\}_{n \geq 1} = H$. Let $X = \text{span}\{e_n\}_{1 \leq n \leq N-1}$ (X is a finite dimensional space, hence it is closed) $Y = \text{cl span}\{e_n\}_{n \geq N}$ so that

$$H = X + Y = \text{cl span } \{e_n\}_{n \geq 1}$$

(in general not a direct sum). We denote $\mathbb{M}$, $\mathbb{M}_X$ and $\mathbb{M}_Y$ the moment operators of $\{e_n\}_{n\in\mathbb{N}}$, $\{e_n\}_{n\leq N-1}$ and of $\{e_n\}_{n\geq N}$. Note that dom $\mathbb{M}_X = H$ and $\mathbb{M}_Y$ is defined on Y since $\{e_n\}_{n\geq N}$ is a Riesz sequence and in fact it is defined on H since

$$f = f_Y + g, \quad f_Y \in Y, \ g \perp Y \implies \begin{cases} \mathbb{M}_Y g = 0, \\ \mathbb{M}_Y f = \mathbb{M}_Y f_Y. \end{cases} \tag{3.22}$$

We prove that dim $[\text{im } \mathbb{M}]^\perp < +\infty$. Using (3.22) we see that the elements of im $\mathbb{M}$ are those elements c for which

$$c = \begin{pmatrix} c_X \\ c_Y \end{pmatrix} = \mathbb{M}f = \begin{pmatrix} \mathbb{M}_X f \\ \mathbb{M}_Y f \end{pmatrix} = \begin{pmatrix} \mathbb{M}_X f \\ \mathbb{M}_Y f_Y \end{pmatrix}$$

where $c_X = \{c_n\}_{1\leq n\leq N-1} \in l^2([1, N-1))$ and $c_Y = \{c_n\}_{n\geq N} \in l^2([N, +\infty))$. The vectors c_Y are arbitrary in $l^2([N, +\infty))$ since $\{e_n\}_{n\geq N}$ is a Riesz sequence.

Our assumption is that

$$f_Y = \mathbb{M}_Y^{-1} c_Y$$

so that the elements of the image of $\mathbb{M}$ have the representation

$$\mathbb{M}f = \begin{pmatrix} \mathbb{M}_X (f_Y + g) \\ \mathbb{M}_Y f_Y \end{pmatrix} = \begin{pmatrix} \mathbb{M}_X \mathbb{M}_Y^{-1} c_Y \\ c_Y \end{pmatrix} + \begin{pmatrix} \mathbb{M}_X g \\ 0 \end{pmatrix} \tag{3.23}$$

where $c_Y \in l^2([N, +\infty))$ and $g \in Y^\perp$ are arbitrary.

Let $(\xi, \eta) \in l^2 = l^2([1, N-1)) \oplus l^2([N, +\infty))$ and $(\xi, \eta) \perp$ im $\mathbb{M}$. Then:

$$\left\langle \left((\mathbb{M}_Y^*)^{-1} \mathbb{M}_X^* \xi + \eta \right), c_Y \right\rangle + \langle M_X^* \xi, g \rangle = 0.$$

This equality must hold for every $c_Y \in l^2([N, +\infty))$ and every $g \in Y^\perp$. We take first $g = 0$ and c_Y arbitrary and then $c_Y = 0$ and arbitrary $g \in Y^\perp$. We see that ξ and η satisfy

$$\eta = - \left(\mathbb{M}_Y^* \right)^{-1} \mathbb{M}_X^* \xi, \quad \langle M_X^* \xi, g \rangle = 0 \ \forall g \in Y^\perp.$$

Hence, the codimension of im $\mathbb{M}$ is at most the dimension of the graph of $\left(\mathbb{M}_Y^* \right)^{-1} \mathbb{M}_X^*$, which is finite since dom $M_X^* = l^2([1, N-1])$ has finite dimension.

Now we prove that the image of $\mathbb{M}$ is closed. Let $\{c^k\} \in$ im $\mathbb{M}$ be a convergent sequence, $c^k \to c^0$. We must prove that $c^0 \in$ im $\mathbb{M}$. Using (3.23) we see that

$$c^k = \begin{pmatrix} c_X^k \\ c_Y^k \end{pmatrix} = \begin{pmatrix} \mathbb{M}_X \mathbb{M}_Y^{-1} c_Y^k \\ c_Y^k \end{pmatrix} + \begin{pmatrix} \mathbb{M}_X g^k \\ 0 \end{pmatrix}, \quad c^k \to c^0 = \begin{pmatrix} c_X^0 \\ c_Y^0 \end{pmatrix}.$$

So, $c_Y^k \to c_Y^0$ and $\mathbb{M}_X g^k \to c_X^0 - M_X M_Y^{-1} c_Y^0$. The result follows since $\{g^k\}$ is convergent, $g^k \to g^0$ as the following observations show:

- the operator $\mathbb{M}_X$ is invertible on $Y^\perp$. In fact, if $\mathbb{M}_X g = 0$ then $g \perp e_n$ for $n \le N - 1$. The definition of g is that $g \perp e_n$ for $n \ge N$ and so it must be $g = 0$.
- Hence, $\mathbb{M}_X^{-1}$ exists on its image, which is finite dimensional. So, $\mathbb{M}_X^{-1}$ is continuous and $g^k = \mathbb{M}_X^{-1} \left(c_X^k - M_X M_Y^{-1} c_Y^k \right) \to \mathbb{M}_X^{-1} \left(c_X^0 - M_X M_Y^{-1} c_Y^0 \right)$.

3.4.1 Riesz Sequences of Exponentials in L^2 Spaces

In this section, the scalar field is $\mathbb{C}$.

Riesz sequences in $L^2(0, T; K)$, where K is a Hilbert space, have particular properties. We shall need the following results. See [36] for the first one.

Lemma 3.3 Let $\mathbb{Z}' = \mathbb{Z} \backslash \{0\}$ and let $\{\beta_n\}_{n \in \mathbb{Z}'}$, $\{k_n\}_{n \in \mathbb{Z}'}$ be such that

$$\beta_{-n} = -\beta_n, \quad k_n = k_{-n} \in K, \quad |Im \, \beta_n| < L. \tag{3.24}$$

for e suitable number L. If $\{e^{i\beta_n t} k_n\}_{n \in \mathbb{Z}'}$ is a Riesz sequence in $L^2(0, 2T; K)$, then

$$\{k_n \cos \beta_n t\}_{n > 0}, \quad \{k_n \sin \beta_n t\}_{n > 0} \tag{3.25}$$

are Riesz sequences in $L^2(0, T; K)$.

Proof Thanks to the boundedness of $\{Im \, \beta_n\}$, the transformation

$$\sum \alpha_n e^{i\beta_n t} \mapsto \sum \left(\alpha_n e^{-i\beta_n T} \right) e^{i\beta_n t}$$

is bounded and boundedly invertible from $L^2(0, 2T; K)$ to $L^2(-T, T; K)$. It follows that $\{e^{i\beta_n t} k_n\}_{n \in \mathbb{Z}'}$ is a Riesz sequence in $L^2(-T, T; K)$ and both the inequalities in (3.19) hold in the space $H = L^2(-T, T; K)$. We prove similar inequalities for the sequences (3.25) in $L^2(0, T; K)$. We consider the cosine sequence. Let $\{a_n\}$ be any finite sequence of complex numbers. Then we have

$$\left| \sum_{n>0} a_n k_n \cos \beta_n t \right|^2_{L^2(0,T;K)} = \frac{1}{4} \left| \sum_{n>0} a_n k_n e^{i\beta_n t} + \sum_{n>0} a_n k_n e^{-i\beta_n t} \right|^2_{L^2(0,T;K)}$$

$$= \frac{1}{8} \left| \sum_{n>0} a_n k_n e^{i\beta_n t} + \sum_{n>0} a_n k_n e^{-i\beta_n t} \right|^2_{L^2(-T,T;K)}$$

$$= \frac{1}{8} \left| \sum_{n \in \mathbb{Z}'} a_n k_n e^{i\beta_n t} \right|^2_{L^2(-T,T;K)}.$$

In the last equality, we used we used $-\beta_n = \beta_n, k_{-n} = k_n$ and we have put $a_{-n} = a_n$.

Inequalities (3.19) hold by assumption for the exponential series, hence they hold also for the cosine series.

The proof for the sine sequence is analogous, but put $a_{-n} = -a_n$.

The second result is:

Lemma 3.4 *Let K be a Hilbert space and let $\{k_n\}$ be a sequence in K. Let β_n be a sequence of real numbers such that $\{e^{i\beta_n t}k_n\}$ is a Riesz sequence in $L^2(0, T; K)$. If $\phi(t) = \sum_{n=1}^{+\infty} \alpha_n e^{i\beta_n t}k_n \in H^1(0, T + h_0; K)$ ($h_0 > 0$), then $\alpha_n = (\delta_n/\beta_n)$, $\{\delta_n\} \in l^2$.*

Proof We use this fact ([11, Proposition IX.3] if $K = \mathbb{R}$, easily adapted in general, see problem 3.14): let $\phi \in H^1(0, T + h_0; K)$ and $0 < h < h_0$ then there exists $C = C(\phi) > 0$ independent of h such that

$$\left|\sum_{n=1}^{+\infty} \alpha_n \beta_n \frac{e^{i\beta_n h} - 1}{\beta_n h} e^{i\beta_n t} k_n\right|^2_{L^2(0,T;K)} = \left|\frac{\phi(t + h) - \phi(t)}{h}\right|^2_{L^2(0,T;K)} \leq C. \quad (3.26)$$

Using the fact that $\{e^{i\beta_n t}k_n\}$ is a Riesz sequence in $L^2(0, T; K)$ and the left inequality in (3.19), we see that

$$\sum_{n=1}^{+\infty}\left|\alpha_n \beta_n \frac{e^{i\beta_n h} - 1}{\beta_n h}\right|^2 \leq \frac{1}{m_0}\left|\frac{\phi(t + h) - \phi(t)}{h}\right|^2_{L^2(0,T;K)} \leq C/m_0.$$

The last equality holds for h "small", $|h| < h_0$.

Let s be real. There exists $s_0 > 0$ such that:

$$\left|\frac{e^{is} - 1}{s}\right|^2 = \left(\frac{\cos s - 1}{s}\right)^2 + \left(\frac{\sin s}{s}\right)^2 > \frac{1}{2} \quad \text{for } 0 < s < s_0.$$

Then we have, for every $h \in (0, h_0)$,

$$\sum_{\beta_n < s_0/h} |\alpha_n \beta_n|^2 \leq 2\sum_{n=1}^{+\infty}\left|\alpha_n \beta_n \frac{e^{i\beta_n h} - 1}{\beta_n h}\right|^2 \leq 2\frac{C}{m_0}.$$

The limit for $h \to 0^+$ gives $\{\alpha_n \beta_n\} \in l^2$ as wanted.

To finish this section, we note that when $\{e_n\}$ is a sequence of exponentials then the moment problem is a special interpolation problem (see [98]). In fact, let us consider the moment problem

$$\int_0^T e^{-\lambda_n t} f(t)dt = c_n$$

where f is a (real or complex) function. Then

$$\int_0^T e^{-\lambda_n t} f(t)\mathrm{d}t = \int_0^{+\infty} e^{-\lambda_n t} F(t)\mathrm{d}t = \hat{F}(\lambda_n) \text{ where } F(t) = \begin{cases} f(t) & \text{if } t \in (0, T) \\ 0 & \text{if } t > T. \end{cases}$$

$\hat{F}(\lambda)$ is the Laplace transform of $F(t)$.

A function $F(t) \in L^2(0, +\infty)$ has compact support in $(0, T)$ if and only if

$$\hat{F}(\lambda) \in W_T$$

where W_T is a Hilbert space, whose elements are entire functions of exponential type T and whose restriction to the imaginary axis is square integrable, see [55]. We do not need to enter in more details on this point, but we note that the moment problem (3.2) when $e_n = e^{-\lambda_n t}$ is equivalent to the following interpolation problem: to find a function $\hat{F} \in W_T$ such that

$$\hat{F}(\lambda_n) = c_n.$$

This equivalence of a moment problem and an interpolation problem has been exploited for example in [40, 47, 48, 87].

3.4.2 A Second Worked Example

Following [75, 77], we apply the perturbation Theorems 3.9 and 3.10 in order to prove that a certain sequence is a Riesz sequence. The ideas which will be set forth in this example are the key ideas needed in the application of moment methods to the study of controllability of systems with persistent memory. We present the key ideas in the simplest case

$$N(t) \in C^\infty, \qquad N(0) = 1, \qquad N'(0) = 0, \qquad N''(0) = 0.$$

We study the properties of the sequence $\{z_n(t)\}$ where $z_n(t)$ solves

$$z'_n = -n^2 \int_0^t N(t-s)z_n(s)\mathrm{d}s, \qquad z_n(0) = 1. \tag{3.27}$$

Our goal is to see that $\{z_n(t)\}$ is a Riesz sequence in $L^2(0, T)$, for every $T > \pi$. The derivative of both the sides of (3.27) gives

$$z_n'' = -n^2 z_n - n^2 \int_0^t N'(t-s)z_n(s)ds, \qquad z_n(0) = 1, \quad z_n'(0) = 0$$

from which we get

$$z_n(t) = \cos nt - n \int_0^t \sin ns \left[\int_0^{t-s} N'(t-s-r)z_n(r)dr \right] ds.$$

We integrate by parts twice and we get

$$z_n(t) = \cos nt - \int_0^t N'(t-s)z_n(s)ds$$

$$+ \frac{1}{n} \int_0^t \sin n(t-s) \int_0^s N^{(3)}(s-r)z_n(r)dr\, ds. \qquad (3.28)$$

(we used $N(0) = 1$ and $N'(0) = N''(0) = 0$ in the integration by parts).
Let $L(t)$ be the resolvent kernel of $N'(t)$, defined by

$$L(t) + \int_0^t N'(t-s)L(s)ds = N'(t) \quad \text{so that} \quad L(t) \in C^\infty, \ L(0) = 0, \ L'(0) = 0.$$
$$(3.29)$$

We use (3.28), the properties of Volterra integral equations and we integrate by parts:

$$z_n(t) = \cos nt - \frac{1}{n} \int_0^t L'(t-s) \sin ns\, ds + \frac{1}{n} \int_0^t \sin ns \int_0^{t-s} N^{(3)}(t-s-r)z_n(r)dr\, ds$$

$$- \frac{1}{n} \int_0^t L(t-\tau) \int_0^\tau \sin ns \int_0^{\tau-s} N^{(3)}(\tau-s-r)z_n(r)dr\, ds\, d\tau. \qquad (3.30)$$

Using Gronwall inequality (see Problem 5.1):

Lemma 3.5 *The sequence $\{z_n(t)\}$ is bounded on every interval $[0, T]$. So, for every $T > 0$ there exists $M = M_T$ such that*

$$|z_n(t) - \cos nt| \leq \frac{M}{n}. \qquad (3.31)$$

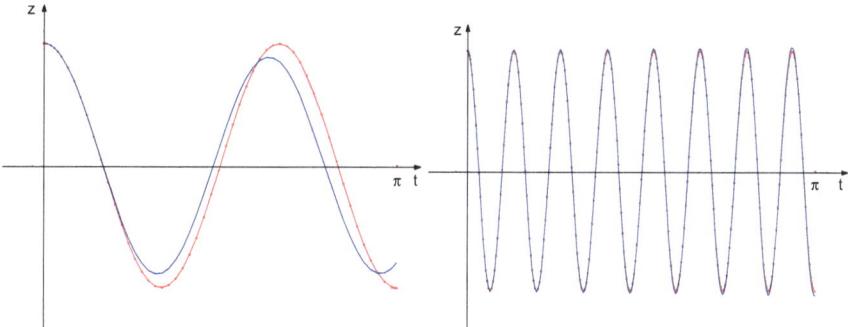

Fig. 3.1 On the interval $[0, \pi]$: $z_n(t)$ with $n = 3$ (*left*) and $n = 15$ (*right*) compared with $\cos 3t$ and $\cos 15t$ (*dotted*)

Remark 3.4 The estimate in the lemma is asymptotic, and improves with n. Figure 3.1 shows this when the kernel is

$$N(t) = 3e^{-t} - 3e^{-2t} + e^{-3t}. \tag{3.32}$$

In this case, the functions $z_n(t)$ are easily computed using the argument in [94] (see Sect. 2.5) since $z_n(t)$ is the first component of the solution (z, w_1, w_2, w_3) of the following problem:

$$\begin{cases} z' = -n^2(3w_1 - 3w_2 + w_3) \\ w_1' = -w_1 + z \\ w_2' = -2w_2 + z \\ w_3' = -3w_3 + z \end{cases} \qquad \begin{pmatrix} z(0) \\ w_1(0) \\ w_2(0) \\ w_3(0) \end{pmatrix} = \begin{pmatrix} 1 \\ 0 \\ 0 \\ 0 \end{pmatrix}.$$

Now, we use the fact that

$$\sum_{n=1}^{+\infty} \frac{1}{n^2} < +\infty$$

and that $\{\cos nt\}$ is an orthogonal sequence in $L^2(0, \pi)$ (elements have constant norms) and a Riesz sequence in $L^2(0, T)$ if $T \geq \pi$. So, we can state (using property 3 in Theorem 3.5 and Theorem 3.9):

Lemma 3.6 *The following hold:*

1. *for every $T > 0$, the sequence $\{z_n(t)\}$ is bounded and quadratically close to $\{\cos nt\}$ in $L^2(0, T)$;*
2. *if $T > \pi$ then there exists a number $N = N_T$ such that $\{z_n(t)\}_{n \geq N}$ is a Riesz sequence in $L^2(0, T)$.*
3. *the series $\sum \alpha_n z_n(t)$ converges in $L^2(0, T)$ if and only if $\{\alpha_n\} \in l^2$.*

In order to prove that the sequence $\{z_n\}$ is Riesz in $L^2(0, T)$ (with $T > \pi$) it is sufficient to prove that it is ω-independent. So, we fix any $T > \pi$ and we prove:

if $\sum_{n=1}^{+\infty} \alpha_n z_n(t) = 0$ (convergence in $L^2(0, T)$)   then $\{\alpha_n\} = 0$.

The proof is in three steps. The first step proves that $\{z_n(t)\}$ is linearly independent and holds for every $T > 0$. The second and third steps instead use explicitly the fact that $T > \pi$.

Step A: the sequence $\{z_n\}$ is linearly independent in $L^2(0, T)$ for every $T > 0$.

If not, there exists a *smaller* K such that the functions z_1, z_2, $\ldots$, z_{K-1}, z_K are *linearly dependent.* It must be $K > 1$ since $z_1(t) \neq 0$. Due to the *minimality* of the choice of K, if we remove z_K we get a *linearly independent* set.

Let $\{\alpha_n\}_{1 \leq n \leq K}$ be such that

$$\sum_{n=1}^{K} \alpha_n z_n(t) = 0. \tag{3.33}$$

This sum must have at least two nonzero coefficients α_n since every function $z_n(t)$ is different from zero.

We compute the derivatives of both the sides of (3.33). We get:

$$0 = \sum_{n=1}^{K} \alpha_n z_n'(t) = -\int_0^t N(t-s) \left[\sum_{n=1}^{n} n^2 \alpha_n z_n(s) \right] ds.$$

The bracket has to be zero since N is differentiable and $N(0) = 1$:

$$\sum_{n=1}^{K} \alpha_n z_n(t) = 0 \implies \sum_{n=1}^{K} n^2 \alpha_n z_n(t) = 0.$$

We multiply the first sum with K^2 and we subtract from the second one. We find

$$\sum_{n=1}^{K-1} \left(K^2 - n^2 \right) \alpha_n z_n(t) = 0$$

which contradicts the minimality of K. This proves linear independence.

Now, we prove ω-independence, again arguing by contradiction. Let

$$\sum_{n=1}^{+\infty} \alpha_n z_n(t) = 0 \text{ in } L^2(0, T). \tag{3.34}$$

We wish to compute the derivative of the series, so to have a new series, still equal zero; and then to combine both the series as we did for a finite sum. The difficulty is that if $\{\alpha_n\}$ is merely of class l^2, the derivative cannot be computed termwise. So, we need an intermediate step, in which we prove that $\{\alpha_n\}$ has stronger decaying properties.

Step B: the decaying properties of $\{\alpha_n\}$, when $T > \pi$.
Using the representation (3.30), condition (3.34) is written as

$$
\sum_{n=1}^{+\infty} \alpha_n \cos nt = \int_0^t L'(t-s) \left[\sum_{n=1}^{+\infty} \frac{\alpha_n}{n} \sin ns \right] ds
$$

$$
- \int_0^t \left(\sum_{n=1}^{+\infty} \frac{\alpha_n}{n} \left[\int_0^{t-\tau} N^{(3)}(t-\tau-r) \sin nr \, dr \right] z_n(\tau) \right) d\tau
$$

$$
+ \int_0^t L(t-s) \left[\sum_{n=1}^{+\infty} \frac{\alpha_n}{n} \int_0^s H_n(s-\tau) z_n(\tau) d\tau \right] ds \qquad (3.35)
$$

where

$$
H_n(\tau) = \int_0^\tau N^{(3)}(\tau-s) \sin ns \, ds .
$$

Note that it is legitimate to distribute the series since all the single series converge in $L^2(0, T)$ when $\{\alpha_n\} \in l^2$. In fact, the numerical series $\sum_{n=1}^{+\infty} \alpha_n/n$ is convergent.

The conditions we imposed on the kernel $N(t)$ show that the right hand side is differentiable with L^2 (in fact continuous) derivative so that

$$
\sum_{n=1}^{+\infty} \alpha_n \cos nt \in H^1(0, T) .
$$

Known properties of the Fourier series (or an application of Lemma 3.4) show that :

$$
\alpha_n = \frac{\delta_n}{n}, \quad \{\delta_n\} \in l^2 .
$$

We replace this expression in (3.35) and we compute the derivative. The condition $\sum_{n=1}^{+\infty} (\delta_n/n) < +\infty$ shows that

$$
\sum_{n=1}^{+\infty} \delta_n \sin nt \in H^1(0, T) \quad \text{so that} \quad \delta_n = \frac{\gamma_n}{n} \quad \text{i.e.,} \quad \alpha_n = \frac{\gamma_n}{n^2} \quad \{\gamma_n\} \in l^2 . \quad (3.36)
$$

Step C: ω-independence in $L^2(0, T)$, when $T > \pi$.

We proved that (3.34) is in fact:

$$\sum_{n=1}^{+\infty} \frac{\gamma_n}{n^2} z_n(t) = 0 \quad \{\gamma_n\} \in l^2 . \tag{3.37}$$

This series can be differentiated termwise. In fact, a formal termwise differentiation gives the convergent series

$$\int_0^t N(t-s) \left[\sum_{n=1}^{+\infty} \gamma_n z_n(s) \right] ds \quad \text{so that we have} \quad \int_0^t N(t-s) \left[\sum_{n=1}^{+\infty} \gamma_n z_n(s) \right] ds = 0.$$

The facts that $N(t)$ is differentiable and $N(0) = 1$ imply

$$\sum_{n=1}^{+\infty} \gamma_n z_n(t) = 0 . \tag{3.38}$$

Combining with (3.37) we get

$$\sum_{n>1} \left(1 - \frac{1}{n^2} \right) \gamma_n z_n(t) = 0 . \tag{3.39}$$

Note that the sequence of the coefficients is in l^2 and that for $n > 1$ we have $\left(1 - (1/n^2) \right) \gamma_n = 0$ if and only if $\alpha_n = 0$.

In conclusion, we found a series whose sum is 0, with l^2 coefficients, similar to (3.34), but now the first index is 2 instead of 1: we peeled off the first element of the series (3.34). The coefficients are changed, but *if a coefficient α_n with $n > 1$ is different from zero, the corresponding coefficient in (3.39) is still different from zero.*

Thanks to the fact that $\{(1 - 1/n^2)\gamma_n\} \in l^2$, the procedure can be iterated till we get

$$\sum_{n \geq N} \tilde{\gamma}_n z_n(t) = 0,$$

which implies $\tilde{\gamma}_n = 0$ for $n > N$ since $\{z_n(t)\}_{n \geq N}$ is a Riesz sequence in $L^2(0, T)$ (see lemma 3.6). So we have also $\alpha_n = 0$ for $n \geq N$ and the series (3.34) is a *finite sum:*

$$\sum_{n=1}^{N} \alpha_n z_n(t) = 0 .$$

Linear independence (proved in the Step A) shows that $\alpha_n = 0$ also for $n \leq N$ and so the sequence $\{z_n(t)\}$ is ω-independent; hence it is a Riesz sequence in $L^2(0, T)$ for every $T > \pi$, as we wanted to prove.

3.5 Problems[4] to Chap. 3

3.1. Let $\mathbb{J} = \mathbb{N}$ and consider the sequence $\{e_n\}$ in $L^2((0, \pi) \times (0, \pi))$ given by $e_n(t, y) = \sin nt \sin y$. Prove that the moment operator is surjective.

3.2. The index set is $\mathbb{N} \times \mathbb{N}$ ordered lexicographically and

$$e_{n,m} = \sin nt \sin my \in L^2((0, \pi) \times (0, \pi)).$$

Prove that the moment operator is surjective.

Note that the function e_n of problem 3.1 is $e_{n,1}$. In spite of this, both the moment operators are surjective.

3.3. The index set is $\mathbb{N} \times \mathbb{N}$ ordered lexicographically and

$$e_{n,m} = \frac{n}{n^2 + m^2} \sin nt \sin y \in L^2((0, \pi) \times (0, \pi))$$

(note: $\sin y$ in the expression of $e_{n,m} = e_{n,m}(t, y)$ is correct). Prove that the moment operator *is not* surjective.

3.4. Let $\{x_n\}$ be an orthonormal basis of H so that any $f \in H$ has the representation $f = \sum_{n=1}^{+\infty} c_n x_n$ and $\{c_n\} \in l^2$. Let $e_n^0 = \sqrt{n} x_n$ Prove that any $f \in H$ can be represented as $f = \sum_{n=1}^{+\infty} f_n e_n^0$ and that the series converges in H. Determine the coefficients $\{f_n\}$ and note that $\{f_n\} \in l^2$.

Let instead $e_n^1 = \frac{1}{\sqrt{n}} x_n$. Prove that any $f \in H$ can be represent as $f = \sum_{n=1}^{+\infty} f_n e_n^1$ but that in general $\{f_n\} \notin l^2$.

Study the moment operators of the two sequences.

When any $f \in H$ has a representation $f = \sum_{n=1}^{+\infty} f_n e_n$ (convergence in H) the sequence $\{e_n\}$ is called a SCHAUDER BASIS of H.

3.5. A theorem of Weierstrass asserts that the set $\{x^n\}_{n \geq 0}$ is dense in $L^2(0, 1)$. Show that not every $f \in L^2(0, 1)$ can be represented as $\sum_{n=0}^{+\infty} \alpha_n x^n$ with $\{\alpha_n\} \in l^2$ (study the regularity of the restriction to $(0, 1/2)$ of the series).

3.6. Use the theorem of Weierstrass mentioned in Problem 3.5 to deduce that $\{1/(x + n)\}_{n \geq 0}$ is complete in $L^2(0, 1)$, using this idea: if the sequence is not complete then there exists $f \in L^2(0, 1)$ such that

---

[4] Solutions at the address http://calvino.polito.it/~lucipan/materiale_html/P-S-CH-3.

$$0 = \int_0^1 f(s)\frac{1}{s+n}ds = \frac{1}{1+n}\int_0^1 f(s)ds + \int_0^1 \left(\int_0^x f(s)ds \right) \frac{1}{(x+n)^2}dx . \quad (3.40)$$

Estimate the last integral and conclude that

$$\left| \frac{1}{1+n}\int_0^1 f(s)ds \right| \le \frac{M}{n^2} \implies \int_0^1 f(s)ds = 0 .$$

Hence, $f \perp 1$. Iterate and prove that $f \perp x^n$ for every n.

Show that the moment operator of $\{1/(x+n)\}_{n\in\mathbb{N}}$ is continuous but not surjective.

3.7. Let $\{x_n\}_{n\ge 0}$ be an orthonormal basis of a Hilbert space H so that any $f \in H$ has the representation $f = \sum_{n=0}^{+\infty} f_n x_n$ with $\sum_{n=0}^{+\infty} |f_n|^2 < +\infty$.
Let $\mathbb{J} = \mathbb{N} = \{1, 2, \ldots\}$ and

$$e_n = x_n - x_0 .$$

If $f \perp e_n$ for every n, then we must have

$$f_n - f_0 = 0 \quad \text{i.e., } \{f_n\} \text{ is constant} .$$

Deduce that $f = 0$ and that $H_0 = H$.

We have $f \in \text{dom } \mathbb{M}$ when $\{\langle f, e_n \rangle\} \in l^2$. Deduce that this is the case if and only if $f_0 = 0$, i.e.,

$$\text{dom } \mathbb{M} = \{x_0\}^\perp ,$$

not dense in $H = H_0$. Study whether $\mathbb{M}$ is surjective and whether it is continuous.

3.8. The definition of the Riesz sequences does depend on the index set $\mathbb{J}$. Let $H = L^2((0, \pi) \times (0, \pi))$. Prove that the sequence $\{e_n(t, y)\}$ is Riesz when the index set is $\mathbb{N}$ and $e_n(t, y) = \sin nt$. Instead, when the index set is $\mathbb{N} \times \mathbb{N}$, then the sequence $\{e_{n,m}(t, y)\}$, $e_{n,m}(t, y) = \sin nt$, is not a Riesz sequence.

3.9 We noted that the left inequality in (3.19) implies that a Riesz sequence is ω-independent. Prove this fact using the properties of the moment operator.

3.10 Let $c > 0$ be not an integer and consider the sequence $\{e_n\}$ in $L^2(0, \pi)$ whose elements are $\cos ct$ and $\cos nt$, $n \ge 1$. It is quadratically close to the Riesz sequence $\{\cos nt\}_{n\ge 0}$. Prove that if $\{e_n\}$ is not ω-independent, then it must be

$$\cos ct = \sum_{n=1}^{+\infty} \frac{\alpha_n}{n^2} \cos nt , \qquad \{\alpha_n\} \in l^2 . \qquad (3.41)$$

Hence:

$$\cos ct = \sum_{n=1}^{+\infty} \alpha_n \cos nt \implies \sum_{n=1}^{+\infty} \alpha_n \left(\frac{1}{n^2} - \frac{1}{c^2} \right) \cos nt = 0 . \qquad (3.42)$$

Deduce that $\alpha_n = 0$ for every n (also for $n = 1$) and deduce that $\{e_n\}$ is a Riesz sequence.

3.11 Note that the sequence whose elements are e^{-ict}, e^{ict} and e^{int} (c not an integer and n any nonzero integer) *is not* a Riesz sequence in $L^2(-\pi, \pi)$. Use this observation and Problem 3.10 to show that the implication in Lemma 3.3 cannot be reversed.
    Decide whether an analogous argument can be applied to the sine series.

3.12 Let $\{e_n(t)\}$ be a *Riesz basis* of $L^2(0, T)$ and let $T_0 \in (0, T)$. we consider the restrictions of $e_n(t)$ to $(0, T_0)$, still denoted $e_n(t)$. Show that the sequence $\{e_n(t)\}$ in $L^2(0, T_0)$ is complete, but that it is not a Riesz sequence.
    Let $\mathbb{M}_{T_0}$ be the moment operator of $\{e_n(t)\}$ in $L^2(0, T_0)$. Show that $\mathbb{M}_{T_0}$ is continuous and compute $\ker \mathbb{M}_{T_0}$.

3.13 Let $\mathbb{M}_{T_0}$ be the moment operator introduced in Problem 3.12. Compute $\mathbb{M}_{T_0}^*$ and its kernel. Conclude that $\left(\mathrm{im}\, \mathbb{M}_{T_0} \right)^{\perp}$ is infinite dimensional.

3.14 Prove inequality (3.26). Discuss the role of the condition $h_0 > 0$.

3.15 Prove that $\{z_n\}$ is a *Riesz basis* if the conditions in Theorem 3.10 hold and furthermore $\{e_n\}$ is a *Riesz basis* of H. This is BARI THEOREM.

Chapter 4
Controllability of the Wave Equation

4.1 Introduction and the Goal of This Chapter

Results on controllability of systems with persistent memory have been derived appealing to the corresponding results of the (memoryless) wave equation. For this reason, in this short chapter, we review the key results on the controllability of wave type equations. First, we describe a "hidden regularity" of the normal derivative of the solutions of the uncontrolled system and the observation inequality. A consequence is that the "active part" Γ of the boundary must be "large" also in terms of the trace of the eigenvectors of A on Γ. Finally, we characterize exact controllability in terms of suitable Riesz sequences.

The ASSOCIATED WAVE EQUATION (shortly, "wave equation") to both the systems with memory (2.1) and (2.3) is either the equation

$$u'' = 2\alpha u' + \Delta u \tag{4.1}$$

or the equation

$$u'' = \alpha^2 u + \Delta u \tag{4.2}$$

with the initial and boundary conditions

$$u(0) = 0, \ u'(0) = 0, \qquad u = f \text{ on } \Gamma, \quad u = 0 \text{ on } \partial\Omega \setminus \Gamma.$$

Note that we have put the velocity term $c_0 = 1$. This is not restrictive, see Sect. 2.4. Here, $u = u(x, t)$ is real valued and, as stated in Sect. 2.1, Ω is a bounded region (on one side of its boundary) with smooth boundary (of class C^2). The case $\Gamma = \partial\Omega$ is not excluded.

Remark 4.1 Note that it is equivalent to study the Eq. (4.1) or (4.2). In fact, if u solves (4.1), then $u_\alpha(x, t) = e^{-\alpha t} u(x, t)$ solves (4.2) with control $e^{-\alpha t} f(x, t)$. So, in the study of control problems, we can decide to use either the representation (4.1)

© The Author(s) 2014
L. Pandolfi, *Distributed Systems with Persistent Memory*, SpringerBriefs in Control, Automation and Robotics, DOI 10.1007/978-3-319-12247-2_4

or (4.2), according to convenience. We shall use the representation (4.1) in Chap. 5 and the representation (4.2) in Chap. 6.

Controllability of the Eq. (4.1) or (4.2) is studied in [53] even when $\alpha = \alpha(x)$ and in [90] when $\alpha = \alpha(x, t)$. We are interested in the case that α is constant. Using the results in [90], the arguments in Chaps. 5 and 6 can be extended to more general cases (see [82]).

Equation (4.1) is controllable when every target (ξ, η) with $\xi \in L^2(\Omega)$, $\eta \in H^{-1}(\Omega)$ can be reached by $(u(T), u'(T))$ at a certain time T, under the action of a suitable "steering" control $f \in L^2(0, T; L^2(\Gamma))$, see the definition in Sect. 2.3. The following properties are known:

- if controllability holds, the control time T can be chosen to be independent by the target (see Theorem 2.5).
- if Γ is "too small," then the wave equation *is not controllable* whatever the time T.
- if T is "too small," then the wave equation *is not controllable,* not even if $\Gamma = \Omega$.
- the set of those times at which the wave equation is controllable is either empty or it is a half line.
- there exist Γ (for example $\Gamma = \partial\Omega$) and T such that the wave equation is controllable at time T (and at larger times) with controls acting on Γ. When $\Gamma = \partial\Omega$ and $\alpha = 0$, then we have controllability at any time T larger than $2(\operatorname{diam}\Omega) = 2\left(\sup_{x,x'\in\Omega} |x - x'|\right)$ (see [66, p. 55–60]).

Controllability is proved using multiplier methods in [53, 66, 90]. Geometric conditions for controllability (both necessity conditions and sufficient conditions, which are essentially sharp) are given in [9, 12], in terms of rays of geometrical optics (an interesting analysis of these geometrical conditions is in [14]).

In the following chapters, we prove that the control properties of the wave equation can be lifted to the equation with memory. We present two independent proofs. The first one (in Chap. 5) is based on moment methods while a second proof (which uses the condition $\Gamma = \partial\Omega$) extends to systems with persistent memory the inverse inequality of the wave equation i.e., the "observability" property of the adjoint system (in Chap. 6). In principle, both the approaches can lead to numerical algorithms for the steering control, not yet studied for systems with memory. See [24] for the memoryless case.

Notations

We recall the notations γ_0 and γ_1 for the traces on $\partial\Omega$ of functions and normal derivatives and we recall that we use a Greek letter to denote the solution in the uncontrolled case, i.e., when $f = 0$ (in this case the initial conditions are not zero of course). Note that this is consistent with the notation ϕ_n used for the eigenvectors of the operator A (introduced in Sect. 1.3. See also Sect. 2.1).

The adjoint equations of (4.1) or (4.2) are:

$$\text{of Eq. (4.1)}: \quad \phi'' = -2\alpha\phi' + \Delta\phi + G \tag{4.3}$$

$$\text{of Eq. (4.2)}: \quad \phi'' = \alpha^2\phi + \Delta\phi + G. \tag{4.4}$$

In both the cases, the initial and boundary conditions are,

$$\phi(x,0) = \phi_0(x) \in H_0^1(\Omega), \quad \phi'(x,0) = \phi_1(x) \in L^2(\Omega), \quad \phi = 0 \text{ on } \partial\Omega.$$

The homogeneous condition on the boundary justifies the notation ϕ.

4.2 Hidden Regularity and Observation Inequality

We refer to [53, 66, 90] and [70, Sect. 2.3.4] for the results below.

It is important to know that the trace on the boundary of the solutions of Eqs. (4.3) and (4.4) are square integrable and, when the system is controllable in time T, $|\gamma_1\phi|_{L^2(G_T)}$ is equivalent to the norm of the initial condition (ϕ_0, ϕ_1).

Lemma 4.1 *Let ϕ solve either Eq. (4.3) or (4.4). For every $T > 0$ and for every $\Gamma \subseteq \partial\Omega$ there exists a number $M = M(T, \Gamma)$ such that the following* DIRECT INEQUALITY *holds:*

$$\int_{\Gamma \times [0,T]} |\gamma_1\phi(x,t)|^2 \, d\Gamma \, dt = \int_0^T \int_\Gamma |\gamma_1\phi(x,t)|^2 \, d\Gamma dt \tag{4.5}$$

$$\leq M \left(|\phi_0|^2_{H_0^1(\Omega)} + |\phi_1|^2_{L^2(\Omega)} + |G|^2_{L^1(0,T;L^2(\Gamma))} \right).$$

This inequality shows a HIDDEN REGULARITY of the solutions of the wave equation: $\gamma_1\phi$ belongs to $L^2(\Gamma \times (0,T))$, for every $T > 0$, when the initial conditions belong to $H_0^1(\Omega) \times L^2(\Omega)$ and $G \in L^1(0, T, L^2(\Gamma))$.

When the wave equation is controllable at a certain time T then Γ, the active part of the boundary, has to be "large" also in terms of the properties of the traces $\gamma_1\phi(x)$ on Γ. In fact, controllability in time T is equivalent to surjectivity of the map $f \mapsto (u(T), u'(T))$ from $L^2(\Gamma \times (0,T))$ to $L^2(\Omega) \times H^{-1}(\Omega)$. It is not difficult to see that this is equivalent to the following property of the adjoint equation:

Lemma 4.2 *System (4.1) (equivalently, system (4.2)) is controllable at time T if and only if there exists $m = m(T, \Gamma) > 0$ such that the following inequality holds for every solution ϕ either of Eq. (4.3) or of Eq. (4.4):*

$$m \left(|\phi_0|^2_{H_0^1(\Omega)} + |\phi_1|^2_{L^2(\Omega)} \right) \leq \int_{\Gamma \times (0,T)} |\gamma_1\phi(x,t)|^2 d\Gamma \, dt. \tag{4.6}$$

The equivalence of controllability and inequality (4.6) is purely functional analytic (see [93, Chap. 11] and Sect. 6.2.1). The proof of controllability in [53, 66, 90] (based on multiplier methods) consists in the proof of inequality (4.6).

Inequality (4.6) is called the OBSERVATION INEQUALITY and, as stated in Lemma 4.2, it holds only if the wave equation is controllable at time T. When coupled with the direct inequality, the observation inequality is also called the INVERSE INEQUALITY. Note that the pair of the direct and inverse inequalities can be formulated as follows: *controllability holds at time T if and only if the function*

$$(\phi_0, \phi_1) \mapsto \left[\int_{\Gamma \times (0,T)} |\gamma_1 \phi|^2 d\Gamma \, dt \right]^{1/2}$$

is a norm on $H_0^1(\Omega) \times L^2(\Omega)$, which is equivalent to the usual norm.

Inequalities (4.5) and (4.6) imply[1]:

Theorem 4.1 *Let the wave equation be controllable at time T. For every target $(\xi, \eta) \in L^2(\Omega) \times H^{-1}(\Omega)$ there exists a (unique) steering control of minimal $L^2(\Gamma \times (0, T))$ norm, and this control is a linear and continuous function of (ξ, η).*

4.2.1 Consequences of Controllability on the Eigenvectors

Let $\{\phi_n\}$ be an orthonormal sequence of eigenvectors of the operator A,

$$\phi_n \in H^2(\Omega) \cap H_0^1(\Omega), \qquad A\phi_n = \Delta\phi_n = -\lambda_n^2 \phi_n.$$

We recall $\lambda_n^2 > 0$ (and $\lambda_n = \sqrt{\lambda_n^2} > 0$). We introduce the numbers

$$\beta_n = \sqrt{\lambda_n^2 - \alpha^2}.$$

The properties of the eigenvalues imply that $\beta_n > 0$, for large n. There might be a finite set of indices for which β_n is on the imaginary axis, possibly $\beta_n = 0$.

Controllability implies that Γ has to be "large" also with respect to the properties of the eigenvectors, in the sense that the sequence of the normal derivatives on Γ of the eigenvectors cannot "grow slowly" as we prove now.

First, we prove (see [88, 95] for an improved version):

Lemma 4.3 *Let Eq. (4.1) (i.e., Eq. (4.2)) be controllable at time $T > 0$. Let $A\phi = -\lambda_n^2 \phi$. If $\gamma_1 \phi = 0$ on Γ, then the initial condition is 0 and so also $\phi = 0$.*

Proof The function $\phi(x, t) = \phi(x)e^{-\alpha t} \cos \beta_n t$ (β_n might be zero) solves Eq. (4.3) with $G = 0$ and initial conditions

$$\phi(x, 0) = \phi(x), \qquad \phi'(x, 0) = 0.$$

---

[1] see also the variational approach to controllability in [66, 70].

Thanks to controllability, we can apply the inequality in (4.6) and we see that if $\gamma_1\phi = 0$ on Γ, then the initial condition is zero and so also $\phi = 0$.

In fact, a stronger property holds: the sequence of the normal derivatives of eigen-vectors must grow, of the same order as $\{\lambda_n\}$. Recall that $\{\psi_n\}$ is an ALMOST NORMAL SEQUENCE in a Hilbert space when there exist $m > 0$ and M such that

$$0 < m \le |\psi_n| \le M.$$

Lemma 4.4 *Controllability at a certain time $T > 0$ implies that the sequence $\{\gamma_1\phi_n/\lambda_n\}$ and equivalently the sequence whose elements are*

$$\Psi_n = \frac{\gamma_1\phi_n}{\beta_n} \quad \beta_n \ne 0, \tag{4.7}$$

are almost normal.

Proof We prove the statement for the sequence $\{\Psi_n\}$. We solve Eq. (4.3) with $G = 0$ and initial conditions

$$\phi(x, 0) = 0, \qquad \phi'(x, 0) = \phi_n(x).$$

The solution is

$$\phi(x, t) = \frac{1}{\beta_n} e^{-\alpha t} \phi_n(x) \sin \beta_n t.$$

Using $|\phi_n|_{L^2(\Omega)} = 1$ and the direct and inverse inequalities, we get

$$m \le \int_{\Gamma\times(0,T)} \left(e^{-\alpha t}\sin\beta_n t\right)^2 \left|\frac{\gamma_1\phi_n}{\beta_n}\right|^2 d\Gamma\,dt = \left[\int_0^T \left(e^{-\alpha t}\sin\beta_n t\right)^2 dt\right]\int_\Gamma |\Psi_n|^2 d\Gamma < M.$$

The result follows since

$$\lim_{n\to+\infty}\int_0^T e^{-2\alpha t}\sin^2\beta_n t\,dt = \frac{1}{4\alpha}\left(1 - e^{-2\alpha T}\right) \qquad \text{if } \alpha \ne 0$$

$$\lim_{n\to+\infty}\int_0^T \sin^2\beta_n t\,dt = \frac{1}{2}T \qquad \text{if } \alpha = 0.$$

See [44] for the idea of this proof.

4.3 Controllability and Moment Problem

For the sake of simplicity of exposition, *we assume that* $-\alpha^2$, *is not an eigen-value of the operator A*, i.e., that

$$\beta_n \neq 0$$

for *every* n. See [82] and Remark 4.2 for the general case.

We use the fact that $\{\phi_n\}$ is an orthonormal basis of $L^2(\Omega)$ while $\{\lambda_n\phi_n\}$ and $\{\beta_n\phi_n\}$ are Riesz bases of $H^{-1}(\Omega)$ (see Sect. 1.3) and we represent the solutions of (4.1) as

$$u(x,t) = \sum_{n=1}^{+\infty} \phi_n(x)u_n(t), \quad u'(x,t) = \sum_{n=1}^{+\infty} \phi_n(x)u_n'(t) = \sum_{n=1}^{+\infty} (\beta_n\phi_n(x))\left(\frac{1}{\beta_n}u_n'(t)\right).$$

$$(4.8)$$

Theorem 2.1 shows that:

Lemma 4.5 *For every* $T > 0$, *we have* $\{(u_n(T),(1/\beta_n)u_n'(T))\} \in l^2$ *and depends continuously on the control* $f \in L^2(0,T;L^2(\Gamma))$: *there exists* $M > 0$ *such that*

$$\sum_{n=1}^{+\infty} u_n^2(T) + \sum_{n=1}^{+\infty} \left(\frac{1}{\beta_n}u_n'(T)\right)^2 \leq M|f|^2_{L^2(0,T;L^2(\Gamma))}.$$

Noting that Eq. (4.1) is the same as Eq. (2.1) with $M = 0$, $c_0 = 1$, $c = \alpha$, and $F = 0$ we can adapt formula (2.35). We get

$$u_n''(t) = 2\alpha u_n' - \lambda_n^2 u_n - \int_\Gamma (\gamma_1\phi_n)\, f(x,t)\mathrm{d}\Gamma$$

so that (we recall that we are working in the case $\beta_n \neq 0$ for every n):

$$\begin{cases} u_n(t) = -\int\limits_{\Gamma\times(0,t)} \mathrm{e}^{\alpha s}\left[\frac{\gamma_1\phi_n}{\beta_n}\sin\beta_n s\right] f(x,t-s)\mathrm{d}\Gamma\,\mathrm{d}s, \\[4mm] u_n'(t) = -\int\limits_{\Gamma\times(0,t)} \mathrm{e}^{\alpha s}\,(\gamma_1\phi_n)\left[\frac{\alpha}{\beta_n}\sin\beta_n s + \cos\beta_n s\right] f(x,t-s)\mathrm{d}\Gamma\,\mathrm{d}s. \end{cases} \quad (4.9)$$

We expand the target (ξ, η) as

$$\xi = -\sum_{n=1}^{+\infty} \xi_n \phi_n(x), \qquad \eta = -\sum_{n=1}^{+\infty} \eta_n \left(\beta_n \phi_n(x)\right) \tag{4.10}$$

(the minus sign is for convenience in the following formulas). Controllability of the wave Eq. (4.1) is equivalent to the solvability of the following equations:

$$u_n(T) = -\xi_n, \qquad \frac{1}{\beta_n} u'_n(T) = -\eta_n;$$

i.e., it is equivalent to solvability of the following moment problem:

$$\int_{\Gamma \times (0,T)} e^{\alpha s} \left[\Psi_n \sin \beta_n s\right] f(x, T - s) \mathrm{d}\Gamma \, \mathrm{d}s = \xi_n \tag{4.11}$$

$$\int_{\Gamma \times (0,T)} e^{\alpha s} \left[\Psi_n \left(\frac{\alpha}{\beta_n} \sin \beta_n s + \cos \beta_n s\right)\right] f(x, T - s) \mathrm{d}\Gamma \, \mathrm{d}s = \eta_n \tag{4.12}$$

where $\{\xi_n\}$ and $\{\eta_n\}$ belong to l^2 and $\Psi_n = (\gamma_1 \phi_n)/\beta_n$, as defined in (4.7).

The function f and the numbers ξ_n and η_n are real.

In order to have more compact formulas, it is convenient to give a complex form to this real moment problem. We introduce $c_n = \eta_n + i\xi_n$. Then, $\{c_n\}$ is an arbitrary (complex valued) l^2 sequence and the moment problem (4.11)–(4.12) takes the form

$$\int_{\Gamma \times (0,T)} \Psi_n \left[e^{i\beta_n s} + \frac{\alpha}{\beta_n} \sin \beta_n s\right] \left(e^{\alpha s} f(x, T - s)\right) \mathrm{d}\Gamma \, \mathrm{d}s = c_n, \quad n \in \mathbb{N} \tag{4.13}$$

where f is *real valued*.

So, controllability of the wave equation is equivalent to solvability of the moment problem (4.13). The steering control $f \in L^2(0, T; L^2(\Gamma))$ of minimal norm, which exists thanks to Corollary 4.1, is the solution of minimal norm of the moment problem and conversely.

We recall that $n > 0$. It is convenient to reformulate the moment problem with

$$n \in \mathbb{Z}' = \mathbb{Z} \setminus \{0\}.$$

We proceed as follows: for $n < 0$ we define

$$\beta_n = -\overline{\beta}_{-n}, \quad \phi_n = \phi_{-n}, \quad \lambda_n = -\lambda_{-n}.$$

Then, we consider the moment problem (4.13) with

$$n \in \mathbb{Z}' = \mathbb{Z} \setminus \{0\}, \quad \{c_n\} \in l_2(\mathbb{Z}')$$

but with *complex valued* control f, i.e., we consider

$$\int_{\Gamma \times (0,T)} f(x, T - s) \left[\Psi_n \left(e^{i\beta_n s} + \frac{\alpha}{\beta_n} \sin \beta_n s \right) \right] d\Gamma \, ds = c_n, \quad n \in \mathbb{Z}'. \quad (4.14)$$

We have (the proof is equal to that of the corresponding Lemma 5.1):

Lemma 4.6 *Assume that the moment problem (4.13) has a* real *valued solution for* every *complex valued* $\{c_n\} \in l^2(\mathbb{N})$. *Then, the moment problem (4.14) has a* complex valued *solution for every* $\{c_n\} \in l^2(\mathbb{Z}')$ *and conversely.*

Problem (4.13) has a real *solution which depends continuously on* $\{c_n\} \in l^2(\mathbb{N})$ *if and only if a solution exists for the moment problem (4.14), which depends continuously on* $\{c_n\} \in l^2(\mathbb{Z}')$.

Hence, controllability of the wave equation is equivalent to solvability of the moment problem (4.14), *where* $f \in L^2(\Gamma \times (0, T))$ *and* $\{c_n\}$ *are complex valued.* We combine this observation with Lemma 4.5 and we see that the moment operator of problem (4.14) is defined on $L^2(0, T; L^2(\Gamma))$ and it is continuous. Theorem 4.1 shows that controllability implies the existence of the steering control of minimal norm, which is a continuous function of the target. Using Theorem 3.4, we can state (when $\beta_n \neq 0$, for every n, the case we are studying just for the sake of simplicity):

Theorem 4.2 *The wave equation is controllable at time T with controls acting on Γ if and only if the sequence whose elements are*

$$\Psi_n \left(e^{i\beta_n s} + \frac{\alpha}{\beta_n} \sin \beta_n s \right)$$

is a Riesz sequence in $L^2(\Gamma \times (0, T))$.

Note that $\{\eta_n + i\xi_n - (\gamma/\beta_n)\xi_n\} \in l^2(\mathbb{Z}')$ is as arbitrary as $\{\eta_n + i\xi_n\}$. So:

Theorem 4.3 *System (4.1) is controllable at time T if and only if the sequence*

$$\left\{ \Psi_n \left(e^{i\beta_n s} + \frac{\gamma}{\beta_n} \sin \beta_n s \right) \right\}$$

(γ arbitrary and fixed, possibly $\gamma = 0$) is a Riesz sequence in $L^2(\Gamma \times (0, T))$.

Remark 4.2 For simplicity, we presented the results in the case that $-\alpha^2$ is not an eigenvalue of A, i.e., when $\beta_n \neq 0$, for every n. If $-\alpha^2$ is an eigenvalue, it might not be a simple eigenvalue, so that there is a *finite* set if indices n such that $-\lambda_n^2 = -\alpha^2$. The equations in (4.13) which correspond to $\beta_n = 0$ have to be replaced with

$$\int\limits_{\Gamma\times(0,T)} \left[\gamma_1\phi_n(x)\right](1+\alpha s+is)\left(e^{\alpha s}f(x,T-s)\right)d\Gamma\,ds = c_n\,.$$

Theorem 4.3 can be adapted to this more general setting, see [82].

4.4 Problems[2] to Chap. 4

4.1 Prove that if $\phi_0 \in L^2(\Omega)$ and $\phi_1 \in H^{-1}(\Omega)$, then in general $\gamma_1\phi \notin L^2(0,T;L^2(\Gamma))$. Examine the case of the wave equation on $(0,1)$.

4.2 Let w and ϕ solve $w'' = w_{xx}$, for $x > 0$ and conditions

$$w(x,0) = 0, \ w'(x,0) = 0, \ \ w(0,t) = f(t),$$
$$\phi(x,0) = \phi_0(x) \in \mathcal{D}(0,+\infty), \ \phi'(x,0) = \phi_1(x) \in \mathcal{D}(0,+\infty), \ \ \phi(0,t) = 0.$$

Assume that $f \in C^1(0,+\infty) \cap L^2(0,+\infty)$. Integrate by parts the equality

$$\int\limits_0^T \int\limits_0^{+\infty} \phi(x,T-s)\left[w''(x,s) - w_{xx}(x,s)\right]dx\,ds = 0$$

and show that

$$\int\limits_0^{+\infty} \left[\phi_0(x)w'(x,T) + \phi_1(x)w(x,T)\right]dx = \int\limits_0^T \phi_x(0,T-s)f(s)ds. \qquad (4.15)$$

4.3 We consider the transformation Λ_T in $L^2(0,T)$ given by $\Lambda_T f = (w(T),w'(T))$ $\in L^2(0,+\infty) \times H^{-1}(0,+\infty)$ where w is the solution defined in Problem 4.2 (or Problem 1.2. To be pedantic, the solution is obtained extending f with 0 for $t > T$).

Prove that $\Lambda_T^*(\phi_0,\phi_1) = \phi_x(0,T-s)$ (ϕ is the solution described in Problem 4.2). Using Theorem 1.2 (which holds also when applied to a Banach space and its dual), deduce that Λ_T is surjective if and only if there exists $m_0 > 0$ such that

$$\int\limits_0^T |\phi_x(0,t)|^2 dt \geq m_0\left(|\phi_0|^2_{H_0^1(0,+\infty)} + |\phi_1|^2_{L^2(0,+\infty)}\right). \qquad (4.16)$$

4.4 Using formula 1.4, prove that the inequality (4.16) *does not hold.*

---

[2] Solutions at the address http://calvino.polito.it/~lucipan/materiale_html/P-S-CH-4.

Consider the map $\tilde{\Lambda}_T f = w(x, T) \in L^2(0, T)$ (i.e., consider the restriction of $x \mapsto w(x, T)$ to the interval $(0, T)$). Prove a condition for controllability analogous to (4.16) and, using the formula in Problem 1.1, prove that *this condition is satisfied*. Compare with Remark 1.2.

4.5 Consider the control problem $w'' = w_{xx}$ ($x \in (0, 1)$) with null initial conditions and boundary condition $w(0, t) = f(t)$, $w_x(1, t) = 0$. Let $\Lambda_T f = w(T)$. The adjoint system is $\phi'' = \phi_{xx}$ with conditions $\phi(0, t) = \phi_x(1, t) = 0$. Show that Λ_T is surjective if and only if

$$m_0 |\phi_1|^2_{L^2(0,1)} \leq \int_0^T |\phi_x(0, s)|^2 ds, \qquad m_0 > 0 \qquad (4.17)$$

and a suitable choice of ϕ_0. Explain this choice.

4.6 Consider the string equation

$$\phi'' = \phi_{xx} + F, \qquad \begin{cases} \phi(0) = \phi_0 \in H_0^1(0, 1), \\ \phi'(0) = \phi_1 \in L^2(0, 1), \\ \phi(0, t) = \phi(1, t) = 0. \end{cases} \qquad (4.18)$$

Note that the exterior derivative on $\Gamma = \{0\}$ is $-\phi'(0)$, i.e., $(-1)\phi'(x)$ computed with $x \in \Gamma$. Multiply both the sides of the string equation with $h(x) = 2(1 - x)\phi_x(x, t)$ and prove the direct inequality.

4.7 The proof of the direct inequality in Problem 4.6 depends on the choice of h, which is zero on $\partial\Omega \setminus \Gamma$, while if $x \in \Gamma$ then $h = \gamma_1 \phi$ is the normal derivative.

   Find a function with this property when $\Omega \subseteq \mathbf{R}^2$ is a disk, and Γ is the circumference. Discuss the case when Γ is half of the circumference.

4.8 Let us consider the problem $\phi'' = \Delta\phi$ on Ω with homogeneous boundary condition, $\phi = 0$ on $\partial\Omega$. The ENERGY is $(1/2) \int_\Omega \left(\phi'^2 + |\nabla\phi|^2\right) dx$. Prove:

1. CONSERVATION OF ENERGY, i.e., the equality

$$\frac{1}{2} \int_\Omega \left(\phi'(x, t)^2 + |\nabla\phi(x, t)|^2\right) dx = \frac{1}{2} \int_\Omega \left(\phi_1^2(x) + |\nabla\phi_0(x)|^2\right) dx$$

for every t;
2. the functions $|\phi'|_{L^2(\Omega)}$, $|\phi|_{L^2(\Omega)}$, and $|\nabla\phi|_{L^2(\Omega)}$ are bounded on $(0, +\infty)$;
3. the following equality:

$$\int_0^T \int_\Omega \left(\phi'^2 - |\nabla\phi|^2\right) dx\, dt = \int_\Omega \left(\phi'(T)\phi(T) - \phi_1\phi_0\right) dx. \qquad (4.19)$$

4.9 Let $\phi \in C^1(\Omega)$ and assume that $\nabla\phi(x)$ has a continuous extension to $\partial\Omega$. By definition, $\gamma_1\phi(x) = (\nabla\phi(x)) \cdot \mathbf{n}(x)$ where $\mathbf{n}(x)$ is the (exterior) normal derivative to $\Gamma = \partial\Omega$ at the point $x \in \partial\Omega$. Prove that if $\phi = 0$ on Γ, then we have also $\nabla\phi(x) = (\gamma_1\phi(x))\mathbf{n}(x)$.

4.10 Consider the string Eq. (4.18) with $F = 0$. Multiply both the sides with $(x - 1)\phi_x(x, t)$, use Problem 4.8 and prove the inverse inequality, with $\Gamma = \{0\}$.
   Find a suitable multiplier when $\Gamma = \{1\}$.

4.11 Let $D \subseteq \mathbb{R}^2$ be the disk of radius 1 and center $(0, 0)$. The points of contact of the tangents from $(-2, 0)$ divide the circumference in two connected parts. One of this is the active part Γ of the boundary.
   Give an interpretation of the vector $\mathbf{m}(x, y) = (x + 2, y)$ when (x, y) belongs to the boundary of the disk.
   Prove the inverse inequality for the wave equation in D using the following method: multiply both the sides of the equation with $\mathbf{m}(x, y) \cdot \nabla\phi(x, y)$ where $\mathbf{m}(x, y) = (x + 2, y)$. Use Problem 4.8 and the observation in Problem 4.9.
   This method gives the inverse inequality when the active part of the boundary is one of the two parts. Decide which one.

Chapter 5
Systems with Persistent Memory: Controllability via Moment Methods

5.1 Introduction and the Goal of This Chapter

We study the controllability of the pair (deformation/velocity of deformation) for a viscoelastic body. The idea is that existing controllability results for the *memoryless* wave equation can be lifted to the system with memory. As we have seen, the component *w* can also be interpreted as the temperature of thermodynamic systems with memory, so that we get exact controllability of the temperature (at a suitable time T), a property that cannot hold for the standard heat equation, derived from Fourier law. In this chapter, we show the use of moment methods in the study of controllability of viscoelastic materials, or thermodynamical systems with memory (a different proof based on the observation inequality is in Chap. 6).

The final section shows an application of controllability to a source identification problem.

Controllability[1] if system (2.1) is studied under the action of (real) boundary controls f acting on $\Gamma \subseteq \partial\Omega$. It is convenient to represent the system in the form (2.3) which we rewrite as

$$w'(x,t) = 2\alpha w(x,t) + \int_0^t N(t-s)\Delta w(x,s)\mathrm{d}s, \qquad w(x,0) = 0. \qquad (5.1)$$

Note that the coefficient of $w(t)$ is denoted 2α, as in (2.40), to stress that the transformations described in Sect. 2.4 have been done so that without restriction we can study controllability using

$$N(0) = 1, \qquad N'(0) = 0.$$

---

[1] as defined in Sect. 2.3: the targets belong to $L^2(\Omega) \times H^{-1}(\Omega)$ and we can assume that the affine term and the initial conditions are zero.

© The Author(s) 2014
L. Pandolfi, *Distributed Systems with Persistent Memory*, SpringerBriefs in Control, Automation and Robotics, DOI 10.1007/978-3-319-12247-2_5

So, the ASSOCIATED WAVE EQUATION is

$$u'' = 2\alpha u' + \Delta u. \tag{5.2}$$

The systems (5.1) and (5.2) are controlled on Γ, the active part of the boundary ($\Gamma = \partial\Omega$ is not excluded):

$$w(x, t) = u(x, t) = f(x, t) \ x \in \Gamma \subseteq \partial\Omega, \quad w(x, t) = u(x, t) = 0 \ x \in \partial\Omega \setminus \Gamma.$$

Standing assumptions:

- on the kernel: $N(t) \in C^3(0, T)$ for every $T > 0$;
- on the domain: $\Omega \subseteq \mathbb{R}^d$ is a bounded region (on one side of its boundary) with C^2 boundary and $d \leq 3$ (the case of interest in applications).
- systems (5.1) and (5.2) are controlled on Γ (relatively open in $\partial\Omega$): $w = f$ and $u = f$ on Γ while $w = 0$ and $u = 0$ on $\partial\Omega \setminus \Gamma$.
- there exists a time T_0 at which the wave equation (5.2) is controllable.

We recall that $\{\phi_n\}$ is an orthonormal basis of real eigenvectors of the operator A in $L^2(\Omega)$: $A\phi = \Delta\phi$, dom $A = H^2(\Omega) \cap H_0^1(\Omega)$.
    In order to simplify the exposition, we assume:

- the number $-\alpha^2$ is not an eigenvalue of the operator A. Hence,

$$\beta_n = \sqrt{\lambda_n^2 - \alpha^2} \neq 0, \text{ for every } n \text{ and } \beta_n > 0, \text{ for large } n.$$

We introduce the notation

$$G_t = \Gamma \times (0, t) \quad \text{so that} \quad \int_{G_t} \cdots dG_t = \int_0^t \int_{\Gamma} \cdots d\Gamma \, d\tau$$

and we recall:

$$\Psi_n = \frac{\gamma_1 \phi_n}{\beta_n}, \quad \mathbb{Z}' = \mathbb{Z} \setminus \{0\}. \tag{5.3}$$

See [82] (and Remark 4.2) for the case that α is an eigenvalue of A. Using moment methods, in this chapter we prove the following result:

Theorem 5.1 *Let the stated assumptions hold. Then, Eq. (5.1) (i.e., (2.1)) is controllable at any time $T > T_0$ and for every target $(\xi, \eta) \in L^2(\Omega) \times H^{-1}(\Omega)$ there exists a steering control $f \in L^2(0, T; L^2(\Gamma))$ which depends continuously on the target.*

Combining Theorems 2.7 and 5.1, we have that the systems with and without memory have the same sharp control time.

The proof of Theorem 5.1 uses the representation formulas (2.34) and (2.36):

$$w(x,t) = \sum_{n=1}^{+\infty} \phi_n(x)w_n(t), \qquad w'(x,t) = \sum_{n=1}^{+\infty} \phi_n(x)w'_n(t), \tag{5.4}$$

$$w'_n(t) = 2\alpha w_n(t) - \lambda_n^2 \int_0^t N(t-s)w_n(s)ds$$

$$- \int_0^t N(t-s)\left(\int_\Gamma (\gamma_1\phi_n)\,f(x,s)d\Gamma\right)ds, \qquad w_n(0) = 0. \tag{5.5}$$

5.2 Moment Problem and Controllability of Viscoelastic Systems

Let $z_n(t)$ solve

$$z'_n = 2\alpha z_n - \lambda_n^2 \int_0^t N(t-s)z_n(s)ds, \qquad z_n(0) = 1. \tag{5.6}$$

Using (1.21), we get (in the computation of the derivative, we use $N(0) = 1$):

$$w_n(t) = -\int_0^t z_n(\tau) \int_0^{t-\tau} N(t-\tau-s) \int_\Gamma (\gamma_1\phi_n)f(x,s)d\Gamma\,ds\,d\tau$$

$$= -\int_{G_t} \left\{\left[\int_0^s N(s-\tau)z_n(\tau)d\tau\right](\gamma_1\phi_n)f(x,t-s)dG_t\right\}, \tag{5.7}$$

$$w'_n(t) = -\int_{G_t} \left[z_n(s) + \int_0^s N'(s-\tau)z_n(\tau)d\tau\right](\gamma_1\phi_n)f(x,t-s)dG_t. \tag{5.8}$$

Let $\{-\xi_n\} \in l^2$ and $\{-\eta_n\} \in l^2$ be the sequences of the coefficients of the expansions of the targets $\xi \in L^2(\Omega)$ and $\eta \in H^{-1}(\Omega)$ in series of, respectively, $\{\phi_n\}$ and $\{\lambda_n\phi_n\}$ (see Sect. 1.3). Controllability at time T is equivalent to

$$\int\limits_{G_T} \left((\gamma_1 \phi_n) \int\limits_0^s N(s - \tau) z_n(\tau) d\tau \right) f(x, T - s) dG_T = \xi_n, \qquad (5.9)$$

$$\int\limits_{G_T} \frac{\gamma_1 \phi_n}{\lambda_n} \left(z_n(s) + \int\limits_0^s N'(s - \tau) z_n(\tau) d\tau \right) f(x, T - s) dG_T = \eta_n. \quad (5.10)$$

Problem (5.9) and (5.10) can be written as (Ψ_n is defined in (5.3))

$$\int\limits_{G_T} (Z_n(s) \Psi_n) f(x, T - s) dG_T = c_n = \left(\frac{\lambda_n}{\beta_n} \eta_n + i\xi_n \right) \qquad n > 0 \qquad (5.11)$$

where

$$Z_n(t) = z_n(t) + \int\limits_0^t N'(t - s) z_n(s) ds + i\beta_n \int\limits_0^t N(t - s) z_n(s) ds$$

$$= z_n(t) + \int\limits_0^t K_n(t - s) z_n(s) ds, \quad K_n(t) = N'(t) + i\beta_n N(t). \quad (5.12)$$

Note that $\{c_n\}_{n>0}$ is an arbitrary *complex-valued* l^2 sequence.

Remark 5.1 Comparing with Chap. 3, we see that (5.11) is a moment problem for the sequence $\{\overline{Z}_n \overline{\Psi}_n\}$ which has the same basis properties as $\{Z_n \Psi_n\}$.

We reformulate the *moment problem* (5.11) with $n \in \mathbb{Z}' = \mathbb{Z} \setminus \{0\}$. Let:

$$z_n(t) = z_{-n}(t), \quad \phi_n(x) = \phi_{-n}(x), \quad \lambda_{-n} = -\lambda_n; \quad \beta_{-n} = -\overline{\beta}_n.$$

These equalities imply (note that if β_n is imaginary then $Z_n(t)$ is real)

$$\Psi_{-n} = -\overline{\Psi}_n, \quad Z_{-n}(t) = \overline{Z}_n(t), \quad Z_{-n} \Psi_{-n} = -\overline{Z}_n \overline{\Psi}_n = -\overline{(Z_n \Psi_n)}. \quad (5.13)$$

Then, we have the following result (Lemma 4.6 is a special case):

Lemma 5.1 *Assume that the moment problem (5.11) has a* real-valued *solution for* every *complex-valued* sequence $\{c_n\} \in l^2(\mathbb{N})$. *Then, the moment problem with* $n \in \mathbb{Z}'$

$$\int\limits_{G_T} [Z_n(s) \Psi_n] f(x, T - s) dG_T = d_n \in \mathbb{C}, \quad \{d_n\} \in l^2(\mathbb{Z}') \qquad (5.14)$$

has a (complex-valued) solution for every complex-valued sequence $\{d_n\} \in l^2(\mathbb{Z}')$, *and conversely.*

If the moment problem (5.11) has a (real-valued) solution in $L^2(G_T)$ which depends continuously on $\{c_n\}_{n\in\mathbb{N}}$ then the moment problem (5.14) has a solution in $L^2(G_T)$ which depends continuously on $\{d_n\}_{n\in\mathbb{Z}'}$, and conversely.

Proof In this proof, $((\cdot,\cdot))$ denotes the integral of a product in $L^2(G_T)$. Hence, it is linear in both the arguments. We put $g(x,t) = f(x, T-t)$ so that the moment problems that we must compare can be written as

$$((Z_n\Psi_n, g)) = \eta_n + i\xi_n \quad n \in \mathbb{N} \tag{5.15}$$

$$((Z_n\Psi_n, h+ik)) = r_n + is_n \quad n \in \mathbb{Z}'. \tag{5.16}$$

Note that $g, h, k, \eta_n, \xi_n, r_n$, and s_n denote real quantities.

Let $Z_n\Psi_n = \zeta_n + i\hat{\zeta}_n$ so that, using (5.13):

$$\zeta_{-n} = -\zeta_n, \qquad \hat{\zeta}_{-n} = \hat{\zeta}_n.$$

The moment problem (5.15) takes the following form:

$$((\zeta_n + i\hat{\zeta}_n, g)) = \eta_n + i\xi_n \quad n \in \mathbb{N}, \quad g \text{ real valued.} \tag{5.17}$$

The moment problem (5.16) can be written as

$$\begin{cases} r_n + is_n = ((\zeta_n + i\hat{\zeta}_n, h+ik)) & n \in \mathbb{N}, \\ r_{-n} + is_{-n} = ((\zeta_{-n} + i\hat{\zeta}_{-n}, h+ik)) = ((-\zeta_n + i\hat{\zeta}_n, h+ik)) & n \in \mathbb{N}. \end{cases}$$

We separate real and imaginary parts and we see that the moment problem (5.16) can be reformulated as follows. Note that the index set is $\mathbb{N}$:

$$((\xi_n, h)) - ((\hat{\zeta}_n, k)) = r_n, \qquad ((\zeta_n, k)) + ((\hat{\zeta}_n, h)) = s_n,$$
$$((\zeta_n, h)) + ((\hat{\zeta}_n, k)) = -r_{-n}, \quad ((\zeta_n, k)) - ((\hat{\zeta}_n, h)) = -s_{-n}.$$

This is equivalent to the couple of moment problems (both with arbitrary complex-valued l^2 sequences on the right-hand side)

$$((\zeta_n + i\hat{\zeta}_n, h)) = \frac{1}{2}\{[r_n - r_{-n}] + i[s_n + s_{-n}]\} \quad n \in \mathbb{N}, \quad h \text{ real valued.}$$

$$((\zeta_n + i\hat{\zeta}_n, k)) = \frac{1}{2}\{[s_n - s_{-n}] - i[r_n + r_{-n}]\} \quad n \in \mathbb{N}, \quad k \text{ real valued.}$$

So, the solution of the moment problem (5.16) is the same as the solution of two copies of the moment problem (5.15). This ends the proof.

5.3 The Proof of Controllability

Lemma 5.1 shows that in order to prove Theorem 5.1, we must prove:

Theorem 5.2 *Let the wave equation (5.2) be controllable at time T_0 and $T > T_0$. Then, the sequence $\{Z_n(t)\Psi_n\}_{n\in\mathbb{Z}'}$ is a Riesz sequence in the complex Hilbert space $L^2(G_T)$.*

We break the proof in three steps, which follows the path described in Sect. 3.4.2.

5.3.1 Step A: the Functions $Z_n(t)$ and $R_n(t)$

The function $Z_n(t)$ solves

$$
Z_n' = 2\alpha Z_n - \lambda_n^2 \int_0^t N(t-s)Z_n(s)\mathrm{d}s + K_n(t)\,, \qquad
\begin{cases}
Z_n(0) = 1 \\
K_n(t) = N'(t) + \mathrm{i}\beta_n N(t)\,.
\end{cases}
\tag{5.18}
$$

This is clear because (5.12) is the variation of constants formula of Eq. (5.18) (see Problem 5.3 for a direct proof). Hence, using $N(0) = 1$ and $N'(0) = 0$, we have also:

$$
Z_n'' = 2\alpha Z_n' - \lambda_n^2 Z_n - \lambda_n^2 \int_0^t N'(t-s)Z_n(s)\mathrm{d}s + K_n'(t)\,,
\tag{5.19}
$$

$$
Z_n(0) = 1\,, \quad Z_n'(0) = 2\alpha + \mathrm{i}\beta_n
$$

and

$$
Z_n(t) = \mathrm{e}^{\alpha t}\mathrm{e}^{\mathrm{i}\beta_n t} + \mathrm{e}^{\alpha t}\frac{\alpha}{\beta_n}\sin\beta_n t
$$

$$
+ \frac{1}{\beta_n}\int_0^t \mathrm{e}^{\alpha(t-s)}\sin\beta_n(t-s)\left[K_n'(s) - \lambda_n^2\int_0^s N'(s-r)Z_n(r)\mathrm{d}r\right]\mathrm{d}s\,.
$$

We introduce the slightly simpler functions

$$
R_n(t) = \mathrm{e}^{-\alpha t}Z_n(t)\,.
$$

The function $R_n(t)$ can be used in the place of $Z_n(t)$ since it is clearly equivalent to solve the moment problem (5.14) or the moment problem

$$
\int_{G_T} [R_n(t)\Psi_n]\left(\mathrm{e}^{\alpha s} f(x, T-s)\right)\mathrm{d}G_T = c_n \qquad n\in\mathbb{Z}'.
\tag{5.20}
$$

Let

$$N_1(t) = e^{-\alpha t} N'(t) \text{ so that } N_1(0) = 0.$$

Then, we have

$$R_n(t) = G_n(t) - \frac{\lambda_n^2}{\beta_n} \int_0^t \sin \beta_n (t-s) \int_0^s N_1(s-r) R_n(r) dr\, ds = G_n(t)$$

$$- \frac{\lambda_n^2}{\beta_n^2} \int_0^t \left[N_1(t-s) R_n(s) - \cos \beta_n (t-s) \int_0^s N_1'(s-r) R_n(r) dr \right] ds$$

$$\tag{5.21}$$

where

$$G_n(t) = e^{i\beta_n t} + \frac{\alpha}{\beta_n} \sin \beta_n t + \frac{1}{\beta_n} \int_0^t e^{-\alpha s} N''(s) \sin \beta_n (t-s) ds$$

$$+ i \int_0^t e^{-\alpha s} N'(s) \sin \beta_n (t-s) ds$$

$$= e^{i\beta_n t} + \frac{\alpha}{\beta_n} \sin \beta_n t + \int_0^t N_1(t-s) \left(e^{i\beta_n s} + \frac{\alpha}{\beta_n} \sin \beta_n s \right) ds \tag{5.22}$$

(we integrated by parts the term which contains $N''(t)$ and we used $N'(0) = 0$).

Gronwall inequality (see Problem 5.1) shows:

Lemma 5.2 *For every $T > 0$, there exists $M = M_T$ such that for every n we have:*

$$|R_n(t)| \le M \quad t \in [0, T].$$

Theorem 4.2 shows that the sequence $\left\{ \left(e^{i\beta_n t} + \frac{\alpha}{\beta_n} \sin \beta_n t \right) \Psi_n \right\}$ is a Riesz sequence in $L^2(G_T)$ (if the wave equation is controllable at time T). Furthermore, the linear transformation

$$y \mapsto y(t) + \int_0^t N_1(t-s) y(s) ds$$

is bounded with bounded inverse in $L^2(0, T; L^2(\Gamma))$, for every $T > 0$. So, using property 6 in Theorem 3.5, we get:

Theorem 5.3 *Let the wave equation (5.2) be controllable at time T_0 and so also at larger times. Then, the sequence $\{G_n(t)\Psi_n\}$ is Riesz in $L^2(G_T)$, for every $T \geq T_0$.*

The idea now is that we might extract information on the sequence $\{R_n(t)\Psi_n(x)\}$ comparing it with the Riesz sequence $\{G_n(t)\Psi_n(x)\}$.

5.3.2 Step B: Closeness to a Riesz Sequence

We shall use the following notations:

$$
\begin{cases}
E_n(t) = e^{i\beta_n t}, \quad S_n(t) = \sin \beta_n t, \quad C_n(t) = \cos \beta_n t, \\
\zeta_n = E_n + \frac{\alpha}{\beta_n} S_n = C_n + \left(i + \frac{\alpha}{\beta_n}\right) S_n = E_n + \frac{\alpha}{2\beta_n i} \left(E_n - E_{-n}\right), \\
\tilde{E}_n = \Psi_n E_n, \quad \tilde{S}_n = \Psi_n S_n, \quad \tilde{C}_n = \Psi_n C_n, \quad \tilde{\zeta}_n = \Psi_n \zeta_n, \\
\mu_n = \frac{\lambda_n^2}{\beta_n^2} \quad \text{so that} \ (1 - \mu_n) = -\frac{\alpha^2}{\beta_n^2}.
\end{cases}
\tag{5.23}
$$

The following equalities will be repeatedly used:

$$
\begin{aligned}
(C_n * S_n)(t) &= \frac{t}{2} S_n(t), \\
(C_n * C_n)(t) &= \frac{1}{2} \left(t C_n(t) + \frac{1}{\beta_n} S_n(t) \right), \\
(S_n * S_n)(t) &= \frac{1}{2} \left(\frac{1}{\beta_n} S_n(t) - t C_n(t) \right), \\
(C_n * \zeta_n)(t) &= \frac{t}{2} E_n(t) + \frac{i(\alpha t + 1)}{4\beta_n} \left(E_{-n}(t) - E_n(t) \right), \\
(S_n * \zeta_n)(t) &= \left(\frac{t}{2} + \frac{i}{2\beta_n} + \frac{\alpha}{2\beta_n^2} \right) S_n(t) - \left(\frac{it}{2} + \frac{\alpha t}{2\beta_n} \right) C_n(t).
\end{aligned}
\tag{5.24}
$$

We prove that $\{R_n(t)\Psi_n\}$ is quadratically close to $\{G_n(t)\Psi_n\}$. We rewrite (5.21) as

$$
R_n - \zeta_n + N_1 * (R_n - \zeta_n) = -\frac{\alpha^2}{\beta_n^2} N_1 * R_n + \mu_n N_1' * C_n * R_n.
$$

Let L be the resolvent kernel of N_1 (recall that $N_1(0) = 0$) so that

$$
N_1 * L = N_1 - L, \quad L(0) = 0, \quad N_1' * L = N_1' - L'.
\tag{5.25}
$$

Our assumption that $N \in H^3$ implies that $L''(t) \in L^2(0, T)$, for every $T > 0$. Using both the equalities in (5.25), we get:

$$R_n = \zeta_n - \frac{\alpha^2}{\beta_n{}^2} N_1 * R_n + \mu_n N_1' * C_n * R_n$$

$$+ \frac{\alpha^2}{\beta_n{}^2} (L * N_1) * R_n - \mu_n (L * N_1') * C_n * R_n$$

$$= \zeta_n - \frac{\alpha^2}{\beta_n{}^2} L * R_n + \mu_n L' * C_n * R_n$$

and so

$$R_n = \zeta_n - \frac{\alpha^2}{\beta_n{}^2} L * \zeta_n + \mu_n \left(L' * C_n \right) * \zeta_n + \left(\left[\frac{\alpha^2}{\beta_n{}^2} L - \mu_n L' * C_n \right]^{*2} \right) * R_n \tag{5.26}$$

where the exponent $*2$ denotes iterated convolution.

We note:

$$L' * C_n = \frac{1}{\beta_n} H_n(t), \quad H_n(t) = \left\{ L'(0) \sin \beta_n t + \int_0^t L''(t - s) \sin \beta_n s \, ds \right\}. \tag{5.27}$$

So we have:

Lemma 5.3 *There exists a sequence $\{M_n(t)\}$ of continuous functions defined for $t \geq 0$, bounded on bounded intervals and such that*

$$R_n(t) = E_n(t) + \frac{M_n(t)}{\beta_n} = e^{i\beta_n t} + \frac{M_n(t)}{\beta_n}. \tag{5.28}$$

Now we prove:

Theorem 5.4 *Let us assume that the wave equation (5.3) is controllable at time T. Then, there exists $N > 0$ such that the sequence*

$$\left\{ \left[\zeta_n - \frac{\alpha^2}{\beta_n{}^2} L * \zeta_n + \mu_n L' * (C_n * \zeta_n) \right] \Psi_n \right\}_{|n| > N}$$
$$= \left\{ \tilde{\zeta}_n - \frac{\alpha^2}{\beta_n{}^2} L * \tilde{\zeta}_n + \mu_n L' * \left(C_n * \tilde{\zeta}_n \right) \right\}_{|n| > N} \tag{5.29}$$

is a Riesz sequence in $L^2(G_T) = L^2(0, T; L^2(\Gamma))$.

Proof In this proof L^2 denotes $L^2(G_T) = L^2(0, T; L^2(\Gamma))$.

We use the formulas in (5.24), and $\mu_n - 1 = \alpha^2/\beta_n{}^2$ to rewrite the functions in (5.29) in terms of $\tilde{E}_n$:

$$\tilde{\zeta}_n - \frac{\alpha^2}{\beta_n{}^2} L * \tilde{\zeta}_n + \mu_n L' * \left(C_n * \tilde{\zeta}_n \right) \tag{5.30}$$

$$= \tilde{E}_n + \frac{1}{2} L' * (t\tilde{E}_n) + \left\{ \frac{i}{4\beta_n} \left\{ 2\alpha \left(\tilde{E}_{-n} - \tilde{E}_n \right) + L' * \left[(\alpha t + 1) \left(\tilde{E}_{-n} - \tilde{E}_n \right) \right] \right\} \right.$$

$$+ \frac{\alpha^2}{2\beta_n{}^2} \left[L' * \left(t\tilde{E}_n \right) - 2L * \tilde{E}_n \right]$$

$$\left. + \frac{\alpha^2 i}{4\beta_n{}^3} \left\{ L' * \left[(\alpha t + 1) \left(E_{-n} - E_n \right) \right] + 2\alpha L * \left(E_n - E_{-n} \right) \right\} \right\}. \tag{5.31}$$

Theorem 4.3 with $\gamma = 0$ and controllability of the wave equation implies that $\{E_n \Psi_n\}_{n \in \mathbb{Z}'} = \{\tilde{E}_n\}_{n \in \mathbb{Z}'}$ is a Riesz sequence in L^2.

We construct explicitly a bounded and boundedly invertible transformation $\mathscr{T}$ in L^2, which transforms $\tilde{E}_n$ in the corresponding element of the sequence (5.29), i.e., (5.31), when N is large enough. First, we define a transformation $\mathscr{T}_0$ in L^2:

$$\mathscr{T}_0 y = y + \frac{1}{2} L' * (ty).$$

It is known from the theory of the Volterra integral equation that this transformation is bounded and boundedly invertible in L^2. We define a perturbation $\mathscr{T}_N$ of $\mathscr{T}_0$ as follows. Let

$$X_N = \text{cl span} \{\tilde{E}_n\}_{|n|>N}$$

(N to be determined later on). Any element of X_N has the (unique) representation

$$y = \sum_{|n|>N} \gamma_n \tilde{E}_n, \qquad \{\gamma_n\}_{|n|>N} \in l^2 \tag{5.32}$$

(see Theorem 3.5, items 3 and 5). We define

$$\mathscr{T}_N y = 0 \quad \text{if} \quad y \in X_N^{\perp}$$

and, for $|n| > N$, as the large brace in (5.31), i.e.,

$$\mathscr{T}_N \tilde{E}_n = \frac{i}{4\beta_n} \left\{ 2\alpha \left(\tilde{E}_{-n} - \tilde{E}_n \right) + L' * \left[(\alpha t + 1) \left(\tilde{E}_{-n} - \tilde{E}_n \right) \right] \right\}$$

$$+ \frac{\alpha^2}{2\beta_n{}^2} \left[L' * \left(t\tilde{E}_n \right) - 2L * \tilde{E}_n \right]$$

$$+ \frac{\alpha^2 i}{4\beta_n{}^3} \left\{ L' * \left[(\alpha t + 1) \left(E_{-n} - E_n \right) \right] + 2\alpha L * \left(E_n - E_{-n} \right) \right\}.$$

The operator $\mathscr{T}_N$ is extended by linearity to span $\{\tilde{E}_n\}_{|n|>N}$. It is clear that the extension is bounded and so it can be extended by continuity to X_N. Then, it is extended by linearity and continuity to L^2. The operator $\mathscr{T}$ is

$$\mathscr{T} = \mathscr{T}_0 + \mathscr{T}_N.$$

For $|n| > N$, the function $\mathscr{T}\tilde{E}_n = (\mathscr{T}_0 + \mathscr{T}_N)\tilde{E}_n$ is the function in (5.31), i.e., the n-th element of the sequence (5.29).

We prove that the norm of $\mathscr{T}_N$ can be made as small as we wish, by taking N large enough. In fact, let $y \in X$ be given by (5.32). Then $\mathscr{T}_N y$ is the sum of a finite number of series. One of the series is

$$\frac{i\alpha}{2} \sum_{|n|>N} \frac{\gamma_n}{\beta_n} \tilde{E}_n.$$

Using the fact that $\{\tilde{E}_n\}_{|n|>1}$ is a Riesz sequence in L^2, we see that there exist $m_0 > 0$ and M such that for every $\{\gamma_n\} \in l^2$ (and $y = \sum_{|n|>N} \gamma_n \tilde{E}_n$) we have

$$m_0 \sum_{|n|>N} |\gamma_n|^2 \leq |y|^2_{L^2} \leq M \sum_{|n|>N} |\gamma_n|^2, \qquad |y|^2_{L^2} = \left| \sum_{|n|>N} \gamma_n \tilde{E}_n \right|^2_{L^2}.$$

The numbers m_0 and M do not depend on N. We recall that $\{\beta_n\}$ is increasing (when n is large) and we apply these inequalities when the sequence is $\{\gamma_n/\beta_n\}$. We get

$$\left| \sum_{|n|>N} \frac{\gamma_n}{\beta_n} \tilde{E}_n \right|^2_{L^2} \leq \frac{M}{\beta_N^2} \sum_{|n|>N} |\gamma_n|^2 \leq \frac{M}{m_0 \beta_N^2} |y|^2_{L^2}.$$

The remaining series can be handled analogously and we get the following results: for every $\varepsilon > 0$ there exists N such that $\|\mathscr{T}_N\|_{\mathscr{L}(L^2)} < \varepsilon$.

We use Theorem 1.3: *for every bounded and boundedly invertible operator $\mathscr{T}_0$, there exists a number $\varepsilon = \varepsilon(\mathscr{T}_0) > 0$ such that $\mathscr{T}_0 + \mathscr{T}$ is invertible when $\|\mathscr{T}\| < \varepsilon$.*

Now, the choice of N is clear: we choose N such that

$$\|\mathscr{T}_N\|_{\mathscr{L}(L^2)} < \varepsilon(\mathscr{T}_0).$$

With this choice of N the transformation $\mathscr{T} + \mathscr{T}_N$ is bounded with bounded inverse in L^2, and transforms the Riesz sequence $\{\tilde{E}_n\}_{|n|>N}$ to the sequence in (5.29). This completes the proof.

Now we prove that the sequence $\{\Psi_n R_n\}$ is quadratically close to the Riesz sequence in (5.29). We give the proof in the case dim $\Omega \leq 3$.

We examine the last term in (5.26) (which we multiply with Ψ_n). Using (5.27), Lemma 5.2 and the fact that the sequence $\{\Psi_n\}$ is almost normal (Lemma 4.4) we see that there exists a constant M (which depends on T) such that

$$\left|\left(\left[\frac{\alpha^2}{\beta_n^2}L - \mu_n L' * C_n\right]^{*2}\right) * (\Psi_n R_n)\right|^2_{L^2} \le M \frac{1}{\beta_n^4} . \tag{5.33}$$

Now we use the following asymptotic estimate of the eigenvalues of the laplacian: if dim $\Omega = d$, then there exist $m_0 > 0$ and $m_1 > 0$ such that

$$m_0 \left(n^{2/d}\right) < \lambda_n^2 < m_1 \left(n^{2/d}\right) \tag{5.34}$$

see [71, p. 192] (a hint of the proof is in problems 5.10 and 5.11). This estimate is inherited by $\{\beta_n^2\}$ and so

$$d \le 3 \implies \sum \frac{1}{\beta_n^4} < +\infty.$$

Hence:

Theorem 5.5 *Let* dim $\Omega \le 3$ *and let the wave equation be controllable in time* T_0. We have

1. the sequence $\{\Psi_n R_n(t)\}_{n\in\mathbb{Z}'}$ is quadratically close to the Riesz sequence (5.29);
2. if $T \ge T_0$ there exists $N > 0$ such that $\{R_n(t)\Psi_n\}_{|n|>N}$ is a Riesz sequence in $L^2(G_T)$.

Remark 5.2 The condition $d \le 3$ does not have a crucial role. In fact,

$$\Psi_n R_n = \tilde{\zeta}_n - \frac{\alpha^2}{\beta_n^2}L * \tilde{\zeta}_n + \mu_n \left(L' * Cn\right) * \tilde{\zeta}_n$$

$$+ \left(\left[\frac{\alpha^2}{\beta_n^2}L - \mu_n L' * C_n\right]^{*2}\right) * \left[\tilde{\zeta}_n - \frac{\alpha^2}{\beta_n^2}L * \tilde{\zeta}_n + \mu_n \left(L' * C_n\right) * \tilde{\zeta}_n\right]$$

$$+ \left(\left[\frac{\alpha^2}{\beta_n^2}L\Psi_n - \mu_n L' * C_n\right]^{*4}\right) * (\Psi_n R_n).$$

It is clear that the analogous of Theorem 5.4 holds for the sequence of the functions at the first two lines of the right-hand side, and the square of the L^2-norm of the last term is dominated by $1/\beta_n^8$. Hence, the previous arguments can be applied to dimension $d \le 7$.

The argument can be iterated to every dimension.

Now we invoke Theorem 3.11 and we state:

Theorem 5.6 *Let the wave equation be controllable at time* T_0 *and let* $T \ge T_0$. *The reachable set* $R(T) \subseteq L^2(\Omega) \times H^{-1}(\Omega)$ *of the system with memory (5.1) is closed with finite codimension.*

In the final step, using the perturbation Theorem 3.10, we prove that the sequence $\{R_n(t)\Psi_n\}_{n\in\mathbb{Z}}$ is a Riesz sequence, hence that the reachable set is $L^2(\Omega) \times H^{-1}(\Omega)$. A different proof, based on Theorem 5.6 and observability, is in Chap. 6.

5.3.3 Step C: ω-Independence

Now we prove that the sequence $\{R_n(t)\Psi_n\}$ (hence also $\{Z_n(t)\Psi_n\}$) is ω-independent in $L^2(G_T)$ when the wave equation is controllable at time T_0 and $T > T_0$.

Controllability of the wave equation at time T_0, hence also at any time $T > T_0$, implies convergence in $L^2(G_T)$ $(T \geq T_0)$ of the series

$$\sum_{n\in\mathbb{Z}'} \alpha_n \tilde{E}_n, \quad \sum_{n\in\mathbb{Z}'} \alpha_n \tilde{S}_n, \quad \sum_{n\in\mathbb{Z}'} \alpha_n \tilde{C}_n.$$

This is used to prove that if

$$\sum_{n\in\mathbb{Z}'} \alpha_n R_n(t)\Psi_n = 0 \text{ in } L^2(G_T) \tag{5.35}$$

then $\{\alpha_n\}$ has to decay quite fast and, finally, that it has to be zero. The proof relays on Lemma 3.4 and for this reason we use equality (5.35) on $(0, T)$ with $T > T_0$.

In order to simplify the notations, when the index of the series is not explicitly indicated, we intend that it belongs to $\mathbb{Z}'$.

Using (5.23), (5.26), and (5.27), we rewrite (5.35) in the following form:

$$-\sum \alpha_n \tilde{E}_n = \alpha \sum \frac{\alpha_n}{\beta_n} \tilde{S}_n - \alpha^2 \sum \frac{\alpha_n}{\beta_n^2} L * \tilde{\zeta}_n + \sum \mu_n \frac{\alpha_n}{\beta_n} H_n * \tilde{\zeta}_n$$
$$+ \sum \frac{\alpha_n}{\beta_n^2} \left[\left(\frac{\alpha^2}{\beta_n} L - \mu_n H_n \right)^{*2} \right] * (\Psi_n R_n) \tag{5.36}$$

($\{H_n(t)\}$ is bounded on $[0, T]$). We prove:

Lemma 5.4 *The right-hand side of (5.36) belongs to $H^1(0, T; L^2(\Gamma))$.*

Proof We consider the first series on the right side,

$$\sum \frac{\alpha_n}{\beta_n} \tilde{S}_n. \tag{5.37}$$

We use Theorem 4.3 with $\gamma = 0$ and we see that $\left\{\tilde{E}_n\right\} = \left\{\Psi_n e^{i\beta_n t}\right\}$ is a Riesz sequence in $L^2(G_T)$, for $T \geq T_0$, so that also $\{\tilde{S}_n\}$ and $\{\tilde{C}_n\}$ are Riesz sequences, see Lemma 3.3. Controllability of the wave equation implies L^2-convergence of this series. A formal differentiation gives $\sum \alpha_n \tilde{C}_n$ which is still L^2-convergent.

It follows that the series (5.37) converges to an $H^1(0, T; L^2(\Gamma))$ function. We prove an analogous property of the remaining terms.

The second addendum is clearly differentiable, since it is

$$\int_0^t L(t-s) \left[\sum \frac{\alpha_n}{\beta_n^2} \left(e^{i\beta_n s} + \frac{\alpha}{\beta_n} \sin \beta_n s \right) \Psi_n \right] ds .$$

The third series is

$$\sum \frac{\mu_n \alpha_n}{\beta_n} \left(L'(0) S_n * \tilde{\zeta}_n + L'' * S_n * \tilde{\zeta}_n \right) .$$

Differentiability is seen using the expression of $S_n * \zeta_n$ in (5.24). The last term in (5.36) gives several series. For example, series of the functions

$$\frac{\alpha_n \mu_n^2}{\beta_n^2} \left(L'' * S_n^{*2} + L''^{*2} S_n^{*2} \right) * (\Psi_n R_n) .$$

The series clearly converge since (see (5.24))

$$\sum \frac{\alpha_n \mu_n^2}{\beta_n^2} (S_n * S_n) * (\Psi_n R_n) = \frac{1}{2} \sum \frac{\alpha_n \mu_n^2}{\beta_n^2} \left(\frac{1}{\beta_n} S_n - t C_n \right) * (\Psi_n R_n) .$$

The series is even uniformly convergent (use dim $\Omega \leq 3$). We compute the derivative termwise[2] and we see that the more delicate term is

$$\sum \frac{\alpha_n \mu_n^2}{\beta_n} \int_0^t (t-s) \sin \beta_n (t-s) R_n(s) \Psi_n ds .$$

Convergence of this series follows easily using (5.28).

An analogous procedure can be applied to the remaining terms, and we get $\sum \alpha_n \tilde{E}_n \in H^1([0, T]; L^2(\Gamma))$.

Hence, from Lemma 3.4 we get[3]:

Lemma 5.5 *There exists a sequence* $\{\delta_n\} \in l^2$ *such that* $\alpha_n = \delta_n/\beta_n$.

We replace $\alpha_n = \delta_n/\beta_n$ in (5.36) and we compute the derivatives of both the sides. We get:

---

[2] Note that we take the derivative of $S_n(t)$ and not of $L''(t)$, which would require $N^{(4)}(t)$.

[3] The exponents β_n are not all real, as required in the lemma, but we apply the lemma to $\sum_{n > n_0} \alpha_n \tilde{E}_n$, n_0 sufficiently large, since any finite sum is of class C^1.

$$-i \sum \delta_n \tilde{E}_n = \alpha \sum \frac{\delta_n}{\beta_n} \tilde{C}_n - \alpha^2 \sum \frac{\delta_n}{\beta_n{}^3} L' * \tilde{\zeta}_n$$

$$+ \sum \mu_n \frac{\delta_n}{\beta_n} \left[L'(0)C_n + L''(s) * C_n \right] * \tilde{\zeta}_n$$

$$+ \sum \frac{\delta_n}{\beta_n{}^2} \left[\frac{\alpha^2}{\beta_n{}^2} L' - \mu_n \left(L'(0)C_n + L'' * C_n \right) \right]$$

$$* \left[\frac{\alpha^2}{\beta_n} L - \mu_n H_n \right] * [\Psi_n R_n].$$

Note that we used $L(0) = 0$ and $H_n(0) = 0$ in this computation. It is clear that the series on the right-hand side can be differentiated once more, and that the derivative belongs to $L^2(0, T; L^2(\Gamma))$ (use (5.28) in the computations) so that, invoking Lemma 3.4, we see the existence of $\{\sigma_n\} \in l^2$ such that

$$\delta_n = \frac{\sigma_n}{\beta_n}, \quad \text{i.e.,} \quad \alpha_n = \frac{\sigma_n}{\beta_n{}^2}.$$

It is now easy to see that this procedure can be iterated a third time (without using a third derivative of L) and we get:

Theorem 5.7 *If the wave equation is controllable in time T_0 then*

$$\alpha_n = \frac{\gamma_n}{\beta_n{}^3}, \quad \{\gamma_n\} \in l^2.$$

We recall $R_n(t) = e^{-\alpha t} Z_n(t)$ so that from the Eq. (5.18) of $Z_n(t)$ we see that

$$R'_n(t) = \alpha R_n(t) - \lambda_n^2 \int_0^t N_0(t-s) R_n(s) ds + \hat{K}_n(t) \tag{5.38}$$

where

$$N_0(t) = e^{-\alpha t} N(t), \quad \hat{K}_n(t) = e^{-\alpha t} \left(N'(t) + i\beta_n N(t) \right) = N_1(t) + i\beta_n N_0(t).$$

We use this formula in order to examine the consequences of (5.35) i.e.,

$$\sum \frac{\gamma_n}{\beta_n{}^3} R_n(t)\Psi_n = 0 \quad \text{in } L^2(G_T). \tag{5.39}$$

The series (5.39) converges uniformly so that, computing with $t = 0$, we get:

$$\sum \frac{\gamma_n}{\beta_n{}^3} \Psi_n = 0. \tag{5.40}$$

Hence,

$$\sum \frac{\gamma_n}{\beta_n{}^3} \hat{K}_n(t)\Psi_n = N_1(t)\left[\sum \frac{\gamma_n}{\beta_n{}^3}\Psi_n\right] + iN_0(t)\left[\sum \frac{\gamma_n}{\beta_n{}^2}\Psi_n\right]$$
$$= iN_0(t)\left[\sum \frac{\gamma_n}{\beta_n{}^2}\Psi_n\right].$$

The series converges since $d = \dim \Omega \leq 3$. Furthermore,

$$\sum \frac{\gamma_n}{\beta_n{}^3}\lambda_n^2 \int_0^t N_0(t-s)R_n(s)\Psi_n ds = \int_0^t N_0(t-s)\left[\sum \frac{\gamma_n}{\beta_n{}^3}\lambda_n^2 R_n(s)\Psi_n\right] ds$$

(the last series converges since $\{(\gamma_n\lambda_n^2)/\beta_n{}^3\} \in l^2$). So, using (5.38) and (5.39) we get that *the series* (5.39) *is termwise differentiable.* Hence, we get the equality $\sum(\gamma_n/\beta_n{}^3)R'_n(t)\Psi_n = 0$ and so:

$$\int_0^t N_0(t-s)\sum \frac{\gamma_n\lambda_n^2}{\beta_n{}^3}R_n(s)\Psi_n ds = iN_0(t)\left[\sum \frac{\gamma_n}{\beta_n{}^2}\right]\Psi_n. \tag{5.41}$$

Computing with $t = 0$ and using $N_0(0) = 1$, we see a further property:

$$\sum \frac{\gamma_n}{\beta_n{}^2}\Psi_n = 0 \tag{5.42}$$

and so the right-hand side of (5.41) vanishes. As $N_0(0) = 1$, (5.41) gives

$$\sum \frac{\gamma_n\lambda_n^2}{\beta_n{}^3}R_n(t)\Psi_n = 0. \tag{5.43}$$

We rewrite (5.43) and (5.39) as:

$$\sum \alpha_n R_n(t)\Psi_n = 0, \qquad \sum \alpha_n\lambda_n^2 R_n(t)\Psi_n = 0. \tag{5.44}$$

Let k_1 be an index (of minimal absolute value) for which $\alpha_{k_1} \neq 0$. Combining the equalities in (5.44) we get

$$\sum \left(\alpha_n - \frac{\alpha_n\lambda_n^2}{\lambda_{k_1}^2}\right)R_n(t)\Psi_n = 0.$$

Note that the coefficient of $R_{k_1}(t)\Psi_{k_1}$ is zero.

Let

$$\alpha_n^{(1)} = \left(1 - \frac{\lambda_n^2}{\lambda_{k_1}^2}\right)\alpha_n \quad \text{so that} \quad \begin{cases} \left\{\alpha_n^{(1)}\right\} \in l^2, \\ \alpha_k^{(1)} = 0 \text{ if } \lambda_k = \lambda_{k_1} \\ \text{if } \lambda_k \neq \lambda_{k_1} \text{ then } \alpha_k^{(1)} = 0 \iff \alpha_k = 0. \end{cases}$$

In conclusion, we have an equality similar to (5.39) but in which (at least) the term of index k_1 has been removed:

$$\sum_{n \in \mathbb{Z}' \ \lambda_n \neq \lambda_{k_1}} \alpha_n^{(1)} R_n \Psi_n = 0.$$

Thanks to $\left\{\alpha_n^{(1)}\right\} \in l^2$, we can start a bootstrap argument and repeat this procedure: we find that $\left\{\lambda_n^3 \alpha_n^{(1)}\right\} \in l^2$. We fix a second element k_2 (of minimal absolute value) such that $\alpha_{k_2}^{(1)} \neq 0$ and, using a similar argument, we get

$$\sum_{\substack{n \in \mathbb{Z}' \\ \lambda_n \notin \{\lambda_{k_1}, \lambda_{k_2}\}}} \alpha_n^{(2)} R_n(t)\Psi_n = 0. \tag{5.45}$$

We iterate the procedure and we find, after at most $2N$ iteration of the process:

$$\sum_{|n|>N} \alpha_n^{(N)} R_n(t)\Psi_n = 0 \quad \text{and } \alpha_n \neq 0 \implies \alpha_n^{(N)} \neq 0.$$

If N is large enough, as specified in the second statement of Theorem 5.5, $\{R_n\Psi_n\}_{|n|>N}$ is a Riesz sequence in $L^2(0, T; L^2(\Gamma))$ and so it must be

$$\alpha_n^{(N)} = 0 \text{ when } |n| > N, \text{ hence also } \alpha_n = 0, \text{ when } |n| > N.$$

Consequence: the original equality (5.39) is in fact a finite sum. We rewrite it as

$$\sum_{|n|\leq N} \alpha_n R_n(t)\Psi_n = 0. \tag{5.46}$$

This equality implies $\alpha_n = 0$ for every index since we have the following result. The proof is similar to that of Step A in Sect. 3.4.2, but here we have to take into account that different ϕ_n may have the same eigenvalue and we must invoke Theorem 4.3.

Lemma 5.6 *Let the wave equation (5.2) be controllable at a certain time T_0. Then, the sequence $\{R_n(t)\Psi_n(x)\}$ is linearly independent in $L^2(G_T)$, for every $T > 0$.*

Proof The proof is by contradiction. We recall that the sequence $\{R_n\Psi_n\}$ is linearly dependent when there exist numbers α_n such that

$$\sum_{N \leq n \leq K} \alpha_n R_n \Psi_n = 0 \quad \text{and} \quad \sum_{N \leq n \leq K} |\alpha_n|^2 \neq 0. \tag{5.47}$$

Among all the possible intervals $[N, K]$ of integer numbers for which both these conditions hold, there exists at least one of *minimal* length (which is larger than 1, i.e., $N < K$, since $R_n \Psi_n \neq 0$, for every n). We consider (5.47) where $[N, K]$ is an interval of minimal length, so that $\alpha_K \neq 0$ and the sequence $\{R_n \Psi_n\}_{N \leq n \leq K-1}$ is linearly *independent.* Furthermore, we can consider the sum of only those elements in (5.47) whose coefficient α_n is different from zero:

$$\sum_{\substack{N \leq n \leq K \\ \alpha_n \neq 0}} \alpha_n R_n \Psi_n = 0, \quad \text{and} \quad \sum_{N \leq n \leq K} |\alpha_n|^2 \neq 0. \tag{5.48}$$

We distinguish two cases:

Case 1: all the eigenvalues coincide when $\alpha_n \neq 0$*, i.e.,* $\lambda_n^2 = \lambda^2$ *and so also the functions* $R_n(t)$ *coincide,* $R_n(t) = R(t) \neq 0$*, so that condition (5.48) takes the form*

$$\gamma_1 \Phi = 0, \quad \Phi = \left(\sum_{\substack{N \leq n \leq K \\ \alpha_n \neq 0}} \alpha_n \phi_n \right) \neq 0.$$

The condition $\Phi \neq 0$ follows from the orthogonality of the functions $\phi_n(x)$, because the coefficients α_n are not all zero. So, $\Phi(x)$ *is an eigenvector of the operator* A: the condition $\gamma_1 \Phi = 0$ *cannot hold* if the wave equation is controllable at a certain time T_0, see Theorem 4.3.

Case 2: the eigenvalues are different. In this case at least two indices n_1 and n_2 have the property

$$\lambda_{n_1} \neq \lambda_{n_2}, \quad \alpha_{n_1} \neq 0, \quad \alpha_{n_2} \neq 0.$$

From (5.47), we get in particular

$$\sum_{\substack{N \leq n \leq K \\ u_n \neq 0}} \alpha_n \Psi_n = 0.$$

Equating to zero the derivative of the first sum in (5.47) we get:

$$\int_0^t N_0(t - s) \left[\sum_{\substack{N \leq n \leq K \\ \alpha_n \neq 0}} \lambda_n^2 \alpha_n R_n(s) \Psi_n \right] ds = \sum_{\substack{N \leq n \leq K \\ \alpha_n \neq 0}} \alpha_n \hat{K}_n(t) \Psi_n. \tag{5.49}$$

We use the same procedure as for the series and we see that the right-hand side is

$$iN_0(t) \sum_{\substack{N \le n \le K \\ \alpha_n \ne 0}} \alpha_n \beta_n \Psi_n \quad \text{so that} \quad \sum_{\substack{N \le n \le K \\ \alpha_n \ne 0}} \alpha_n \beta_n \Psi_n = 0$$

(compute with $t = 0$ and use $N_0(0) = 1$). Hence

$$\int_0^t N_0(t-s) \left[\sum_{\substack{N \le n \le K \\ \alpha_n \ne 0}} \lambda_n^2 \alpha_n R_n(s) \Psi_n \right] ds = 0, \quad \text{i.e.,} \quad \sum_{\substack{N \le n \le K \\ \alpha_n \ne 0}} \lambda_n^2 \alpha_n R_n(t) \Psi_n = 0.$$

We are studying the case that *at least two different eigenvalues appear in this sum, i.e., the corresponding coefficients α_n are both nonzero.* Combining with the equality in (5.47), we find

$$0 = \sum_{\substack{N \le n \le K \\ \alpha_n \ne 0}} \alpha_n \left(\lambda_n^2 - \lambda_K^2 \right) R_n \Psi_n = \sum_{\substack{N \le n \le K-1 \\ \alpha_n \ne 0}} \alpha_n \left(\lambda_n^2 - \lambda_K^2 \right) R_n \Psi_n. \tag{5.50}$$

At least one of the coefficients α_n in this sum is not zero and *we get the contradictory statement that* $\{R_n \Psi_n\}_{N \le n \le K-1}$ *is linearly dependent.* This ends the proof.

These arguments complete the proof of Theorem 5.1.

Remark 5.3 Once this controllability property is known for linear systems, a perturbation argument can be used to extend to (mildly) nonlinear systems, as in [5, 6, 62].

5.4 An Application: Source Identification

We outline an application of the results just obtained to a SOURCE IDENTIFICATION problem, from [78]. We consider a thermal system acted upon by a distributed source of heat. Now θ denotes the temperature and the equation is (see Sect. 2.6)

$$\theta' = \int_0^t N(t-s)\Delta\theta(s)ds + b(x)\sigma(t) \tag{5.51}$$

where $b(x)$ represents the source of additional heat and *it is unknown.* For example, it might be the characteristic function of a subregion of Ω. More in general, $b(x) \in L^2(\Omega)$.

In practice, the most important case is when $\sigma(t) \ne 0$ is a known constant. More in general *we assume that $\sigma(t) \in H^1(0, T)$ for every $T > 0$ (so, in particular it is continuous) and that $\sigma(0) \ne 0$.*

An operator can measure the flux of heat from a part Γ of the boundary of Ω and the goal is to use such measures in order to evaluate $b(x)$. We are going to present a method for the computation of its Fourier expansion in series of ϕ_n.

We assume that the initial condition is known, and that the body is kept at known temperature on the boundary, so that it is not restrictive to assume:

$$\theta(x, 0) = 0, \quad \theta = 0 \text{ on } \partial\Omega.$$

We have seen in Sect. 2.6 that the flux on the boundary is

$$q(x, t) = - \int_0^t N(t - s)\gamma_1\theta(x, s)ds \quad \text{(measured on } \Gamma\text{)}.$$

Due to the assumed regularity of $N(t)$ and $N(0) = 1$, from this measured quantity we can compute[4] $y = \gamma_1\theta$ on Γ as the solution of

$$y(t) + \int_0^t N'(t - s)y(s) = q'(t), \quad q'(t) = q'(x, t) \text{ and } y(t) = y(x, t) \text{ in } L^2(\Gamma).$$

We are going to see that the identification problem is solvable if Γ is chosen so to have controllability at a certain time T, and if the flux is observed during an interval of time of duration T. In fact, let $\theta_n(t) = \langle \theta(x, t), \phi_n(x) \rangle$ ($\langle \cdot, \cdot \rangle$ is the inner product in $L^2(\Omega)$). Then

$$\theta'_n = -\lambda_n^2 \int_0^t N(t - s)\theta_n(s)ds + b_n\sigma(t) = -\lambda_n^2 N * \theta_n + b_n\sigma, \quad b_n = \langle \phi_n, b \rangle.$$

So we have

$$\theta_n(t) = b_n z_n * \sigma, \qquad \theta(t) = \sum_{n=1}^{+\infty} \phi_n(x) [b_n(z_n * \sigma)(t)]$$

($z_n(t)$ is defined in (5.6) now with $\alpha = 0$) and so

$$y(x, t) = \sum_{n=1}^{+\infty} (\gamma_1\phi_n(x)) [b_n(z_n * \sigma)(t)].$$

We read the output on the interval of time $[0, T]$ and we keep memory of this information as follows: we fix any function $h(x, t)$ and we compute

---

[4] Efficient algorithms for the numerical computation of derivatives exist, see for example [76].

$$\int_{\Gamma} \int_0^T h(x, T - s) y(x, s) ds \, d\Gamma = \sum_{n=1}^{+\infty} b_n \left(\int_{\Gamma} \int_0^T (\gamma_1 \phi_n) z_n (T - r)(h * \sigma)(r) dr \, d\Gamma \right).$$

We recall the definition of b_n and we see that

$$\int_{\Gamma} \int_0^T h(x, T - s) y(x, s) ds \, d\Gamma$$

$$= \int_{\Omega} b(x) \left[\sum_{n=1}^{+\infty} \phi_n \left(\int_{\Gamma} \int_0^T (\gamma_1 \phi_n) z_n(r)(h * \sigma)(T - r) dr \, d\Gamma \right) \right] dx. \quad (5.52)$$

We compare the bracket with (5.4) and (5.7). We see that the two expressions coincide if we can find h such that

$$\int_0^t h(s) \sigma(t - s) ds = \int_0^t N(t - s) f(s) ds.$$

Such a function h surely exists *for every* $f \in L^2(G_T)$, and it is obtained as the solution of the Volterra integral equation

$$h(t) + \frac{1}{\sigma(0)} \int_0^t \sigma'(t - s) h(s) ds = \frac{1}{\sigma(0)} \left(f(t) + \int_0^t N'(t - s) f(s) ds \right).$$

Thanks to controllability at time T, we can choose h which forces the bracket to take any prescribed value in $L^2(\Omega)$ at time T. In particular, we choose $h = h_k$ such that the bracket in (5.52) is equal to $\phi_k(x)$. With this choice of h_k, we find

$$\int_{G_T} h_k(x, T - s) y(x, s) dG_T = b_k = \int_{\Omega} b(x) \phi_k(x) dx,$$

$$b = \sum_{n=1}^{+\infty} \phi_n \left[\int_{G_T} h_k(T - s) (\gamma_1 \theta) (s) dG_T \right].$$

This is the reconstruction of the unknown source $b(x)$.

Remark 5.4 Note that $\{h_k * \sigma\}$ is biorthogonal to $\{(\gamma_1 \phi_n) z_n\}$.

5.5 Problems[5] to Chap. 5

5.1. Let $u(t)$ be a continuous function. Assume that on $[0, +\infty)$ we have

$$0 \le u(t) \le a + b \int_0^t u(s)ds. \tag{5.53}$$

Prove that for every $T > 0$ there exists M_T such that

$$0 \le u(t) \le (M_T)a.$$

This is a weak version of *Gronwall lemma* which is sufficient for the arguments of this chapter.

5.2. Let $z_n(t)$ and $\zeta_n(t)$ solve the Volterra integrodifferential equations (5.54) below. Use the inequality in Problem 5.1 and prove that the sequences $\{z_n\}$ and $\{\zeta_n\}$ are quadratically close on every interval $[0, T]$. The equations are

$$z_n' = -n^2 \int_0^t z_n(s)ds, \quad z_n(0) = 1,$$

$$\tag{5.54}$$

$$\zeta_n' = -n^2 \int_0^t \left[1 - (t-s)^2\right] \zeta_n(s)ds, \quad \zeta_n(0) = 1.$$

5.3. Let $N(t) \in C^1([0, 2\pi] \times [0, 2\pi])$ (possibly $N(t, t) = 0$). Prove that the sequence $\{g_n(t)\}_{n \in \mathbb{Z}}$ where $g_n(t) = e^{int} + \int_0^t N(t, s)e^{ins}ds$ is a Riesz basis in $L^2(0, 2\pi)$ (use Bari Theorem, i.e., Problem 3.15).

5.4. Let $\lambda_n^2 = n^2$. Consider the functions $z_n(t)$ in (5.6) (for simplicity, with $\alpha = 0$). Prove that the sequence $\left\{n \int_0^t z_n(s)ds\right\}$ is quadratically close to $\{-\sin nt\}$ on every interval $(0, T)$.

5.5. Consider the same sequence $\{z_n(t)\}$ as in Problem 5.4. Let $\varepsilon > 0$. Fix any $K > 0$ and prove the existence of $N = N(\varepsilon, K) > 0$ such that the following inequality holds for any $k \le K$ and any $n \ge N$:

$$\left| \int_0^\pi z_n(t) \cos kt \, dt \right| \le \varepsilon.$$

5.6. Give a direct proof of the integrodifferential equation (5.18) of $Z_n(t)$.

---

[5] Solutions at the address http://calvino.polito.it/~lucipan/materiale_html/P-S-CH-5.

5.7. A kernel $N(t)$ is a POSITIVE REAL KERNEL when it is integrable on $[0, T]$ for every T and furthermore

$$\int_0^t v(\tau) \int_0^\tau N(\tau - s)v(s)ds\, d\tau \geq 0 \qquad (5.55)$$

for every $t \geq 0$ and for every function $v(t) \in C^\infty([0, +\infty))$. prove that if (5.55) holds for every function of class $C^\infty([0, T])$ then it holds also for every v which belongs to $L^2(0, T)$ for every T.

5.8. Let $N(t)$ be a positive real kernel. Prove that if

$$z_n' = -\alpha z_n(t) - \lambda_n^2 \int_0^t N(t - s)z_n(s)ds\,, \qquad z_n(0) = 1$$

with $\lambda_n^2 \geq 0$, $\alpha \geq 0$, then $z_n(t)$ is bounded on $[0, +\infty)$, uniformly respect to n: $|z_n(t)| \leq M$, for every n and every $t \geq 0$. If $\alpha > 0$ and if there exists $\varepsilon \in (0, \alpha)$ such that $e^{\varepsilon t} N(t)$ is positive real then there exists $M > 0$ such that $|z_n(t)| \leq M e^{-\varepsilon t}$, for every $t \geq 0$ and every n.

5.9. Let $N(t) = \sum_{k=0}^{+\infty} a_k e^{-b_k t}$ with $a_k \geq 0$ and $b_k \geq 0$. Prove that $N(t)$ is positive real if the sum is finite and also if $\{a_k\}$ is bounded and $\lim b_k \to +\infty$.

5.10. This and the next problem give a hint on the derivation of the estimate (5.34). Consider the operator A on the square $Q = (0, \pi) \times (0, \pi)$. An orthonormal basis of $L^2(Q)$ whose elements are eigenvalues of A was found in Problem 2.4: the elements are $(4/\pi^2) \sin nx \sin my$, identified by the pairs (n, m) of positive integers. Order these pairs in a sequence as follows: first in increasing order of $R^2 = n^2 + m^2$ and if two pairs (n, m) are on the same circumference, order with increasing m. Let $N(R)$ be the number of the eigenvalues such that $n^2 + m^2 \leq R^2$, i.e., the eigenvalues such that $(n, m) \in B(0, R)$ where $B(0, R)$ is the *closed* disk of center $(0, 0)$ and radius R. Let U_R be the union of the squares of opposite vertices $(n - 1, m - 1)$ and (n, m) with $(n, m) \in B(0, R)$. Compare the areas and give an estimate for $N(R)$.

5.11. We continue with the eigenvalues of the problem 5.10. Let K be the index of one eigenvalue $\lambda = \lambda_K$. Let ν_K be the number of the eigenvalues λ such that $|\lambda| = |\lambda_K|$. Give an estimate of ν_K and prove that $\nu_K \leq 16 |\lambda_K|$. Note that

$$|\lambda_K|^2 - \nu_K \leq N(\lambda_K) \leq |\lambda_K|^2 + \nu_K$$

and use this observation to give an estimate of λ_K / K.

Chapter 6
Systems with Persistent Memory: The Observation Inequality

6.1 Introduction and the Goal of This Chapter

The goal of this chapter is to present a different circle of ideas used in the study of systems with persistent memory. For this, we give a proof of controllability which relays on the direct and inverse inequalities of the wave equation. The key properties we shall encounter are the extension of the hidden regularity to systems with memory, a test for the solutions which shows explicitly the role of the region Ω and propagation of singularities. Furthermore, we prove that the inverse inequality is equivalent to controllability both for systems with and without memory. So, we can conclude that the inverse inequality of the memoryless wave equation is inherited by the system with memory.

The point of departure of this chapter is Theorem 5.6: *if the associated wave equation is controllable at time T_0 and $T \geq T_0$ then the reachable set $R(T) \subseteq L^2(\Omega) \times H^{-1}(\Omega)$ of the system with memory is closed with finite codimension.* The final results in Sect. 6.3 of this chapter use the restrictive condition that the control acts on the entire boundary of Ω, $\Gamma = \partial\Omega$, (not used in Chap. 5).

It is convenient to apply the MacCamy trick followed by the transformations described in Sect. 2.4, i.e., we study controllability of the following system (which is (2.38) with $c_0 = 1$ and $v_0 - \alpha u_0$ renamed v_0):

$$
\begin{cases}
w'' = \Delta w + hw + \displaystyle\int_0^t K(t-r)w(r)\,\mathrm{d}r + F(t) = \Delta w + hw + K*w + F(t) \\
w(0) = u_0, \ w'(0) = v_0, \qquad w = f \text{ on } \Gamma, \ w = 0 \text{ on } \partial\Omega \setminus \Gamma.
\end{cases}
$$
$$(6.1)$$

We recall: 1. when studying controllability we can assume $u_0 = 0$, $v_0 = 0$ and $F = 0$.
2. if u_0, v_0 belong to $\mathscr{D}(\Omega)$, $F \in \mathscr{D}(\Omega \times (0, T))$ and $f \in \mathscr{D}(\Gamma \times (0, T))$ then we have the following equality, which holds for every t (see Eqs. (2.20) and (2.26)):

© The Author(s) 2014
L. Pandolfi, *Distributed Systems with Persistent Memory*, SpringerBriefs in Control, Automation and Robotics, DOI 10.1007/978-3-319-12247-2_6

$$w''(t) = A\left(w(t) - Df(t)\right) + hw(t) + \int_0^t K(t-s)w(s)\,ds + F(t)\,. \qquad (6.2)$$

3. when $f = 0$ we use a Greek letter:

$$\phi'' = \Delta\phi + h\phi + \int_0^t K(t-r)\phi(r)\,dr + F(t), \qquad \begin{cases} \phi(0) = \phi_0,\ \phi'(0) = \phi_1, \\ \phi = 0 \text{ on } \partial\Omega. \end{cases}$$

$$(6.3)$$

6.2 Hidden Regularity for Systems with Persistent Memory, and a Test for the Solutions

The solutions of the equation with persistent memory, and their properties, have been studied in Chap. 2 in terms of cosine operator theory. A shortcoming of this approach is that the region Ω does not appear explicitly in the formulas, hidden in the operator A, hence in the sine and cosine operators. In this section, we present a test for the solutions of Eq. (6.1) which shows explicitly the role of the region Ω. As a first step, we extend to systems with memory the direct inequality of the wave equation, presented in Sect. 4.2.

We recall that the direct inequality shows a special hidden regularity of γ_1 on the solutions of the wave equation, which holds when $F \in L^1(0, T; L^2(\Omega))$, $f = 0$, and initial conditions from $H_0^1(\Omega) \times L^2(\Omega)$.

It is easy to extend the direct inequality to the system (6.3): we write Eq. (6.3) as

$$\psi'' = \Delta\psi + \tilde{F}\,, \quad \tilde{F}(t) = F(t) + h\phi + \int_0^t K(t-s)\phi(s)\,ds\,.$$

$$\psi(0) = \phi_0 \in H_0^1(\Omega)\,, \quad \psi'(0) = \phi_1 \in L^2(\Omega)\,, \quad \psi = 0 \text{ on } \partial\Omega\,.$$

Using Theorem 2.4, we have the existence of $M = M_T$ such that

$$|\tilde{F}|^2_{L^1(0,T;L^2(\Omega))} \leq M\left\{|\phi_0|^2_{H_0^1(\Omega)} + |\phi_1|^2_{L^2(\Omega)} + |F|^2_{L^1(0,T;L^2(\Omega))}\right\}. \qquad (6.4)$$

The direct inequality *of the wave equation* gives

$$|\gamma_1\psi|^2_{L^2(0,T;L^2(\Gamma))} \leq M\left(|\phi_0|^2_{H_0^1(\Omega)} + |\phi_1|^2_{L^2(\Omega)} + |\tilde{F}|^2_{L^1(0,T;L^2(\Omega))}\right).$$

Finally, unicity of the solution of Eq. (6.3) shows that $\psi = \phi$ and so, inserting (6.4) in the previous inequality, we get the DIRECT INEQUALITY *for the system with memory*:

Lemma 6.1 *Let ϕ solve Eq. (6.3) and let $\phi_0 \in H_0^1(\Omega)$, $\phi_1 \in L^2(\Omega)$, $F \in L^1(0, T; L^2(\Omega))$. For every $\Gamma \subseteq \Omega$ and every $T > 0$ there exists $M = M(T, \Gamma)$ such that*

$$|\gamma_1\phi|^2_{L^2(0,T;L^2(\Gamma))} \leq M\left(|\phi_0|^2_{H_0^1(\Omega)} + |\phi_1|^2_{L^2(\Omega)} + |F|^2_{L^1(0,T;L^2(\Omega))}\right). \tag{6.5}$$

Inequality (6.5) is the HIDDEN REGULARITY of γ_1 on the solutions of the system with memory.

We use the direct inequality in order to give a test for the solutions of Eq. (6.1).

As in Sect. 1.3, $\langle\langle\cdot, \cdot\rangle\rangle$ denotes the duality pairing of $H_0^1(\Omega)$ and $H^{-1}(\Omega)$. We recall:

- Let $w \in L^2(\Omega)$. The regular distribution

$$\phi \mapsto \langle\langle\phi, \chi_w\rangle\rangle = \int_\Omega \phi(x)w(x)\,dx \tag{6.6}$$

  belongs to $H^{-1}(\Omega)$.
- if $\phi \in H_0^1(\Omega)$ and if $\chi_{w_n} \to \chi \in H^{-1}(\Omega)$ then we have (see (1.28))

$$\lim\langle\langle\phi, \chi_{w_n}\rangle\rangle = \lim \int_\Omega \phi(x)w_n(x)\,dx = \langle\langle\phi, \chi\rangle\rangle. \tag{6.7}$$

- $\mathscr{D}(\Omega)$ is dense in $L^2(\Omega)$. It is also dense in $H^{-1}(\Omega)$ in the sense that the regular distributions represented as in (6.6) with $w \in \mathscr{D}(\Omega)$ are dense in $H^{-1}(\Omega)$.

Using these properties we prove:

Theorem 6.1 *Let $u_0 \in L^2(\Omega)$, $v_0 \in H^{-1}(\Omega)$, $F \in L^1(0, T; L^2(\Omega))$, $f \in L^2(0, T; L^2(\Gamma))$. The function $w \in C(0, T; L^2(\Omega)) \cap C^1(0, T; H^{-1}(\Omega))$ solves Eq. (6.1) if and only if the following equality holds*

$$\langle\langle\phi_0, w'(t)\rangle\rangle + \int_\Omega \phi_1 w(t)\,dx = -\int_0^t \int_\Gamma f(s)\,(\gamma_1\phi(t-s))\,d\Gamma\,ds$$

$$+ \langle\langle\phi(t), v_0\rangle\rangle + \int_\Omega \phi'(t)u_0\,dx$$

$$+ \int_0^t \int_\Omega \phi(t-\tau)F(\tau)\,dx\,d\tau \tag{6.8}$$

for every solution[1] ϕ *of*

$$\begin{cases} \phi'' = \Delta\phi + h\phi + \int_0^t K(t-r)\phi(r)\,dr, \\ \phi(0) = \phi_0 \in \mathscr{D}(\Omega), \ \phi'(0) = \phi_1 \in \mathscr{D}(\Omega) \, and \, \phi = 0 \, on \, \partial\Omega. \end{cases} \quad (6.9)$$

Proof We prove necessity: we assume that w is a solution of Eq. (6.1) and we prove formula (6.8). We first prove this fact when u_0, u_1, F, and f are of class C^∞ with compact support. In this case, equality (6.2) holds for every t and the following computations make sense (see Theorem 2.4).

We fix $T > 0$ and for each $t < T$ we multiply both the sides of Eq. (6.2) with $\phi(t-\tau)$. Then we integrate on $\Omega \times (0,t)$ (the dependence of the functions on x is not indicated). Integrating by parts the left-hand side we get

$$\int_\Omega \int_0^t \phi(t-\tau)w''(\tau)\,d\tau\,dx = \int_\Omega \left[\phi_0 w'(t) - \phi(t)v_0 + \phi_1 w(t) - \phi'(t)u_0\right]dx$$

$$+ \int_\Omega \int_0^t \phi''(t-\tau)w(\tau)\,d\tau\,dx. \quad (6.10)$$

We integrate by parts the first term on the right-hand side, using $\phi(t) \in \text{dom } A$ and (2.14):

$$\int_0^t \int_\Omega \phi(t-\tau)A\left[w(\tau) - Df(\tau)\right]dx\,d\tau$$

$$= \int_0^t \int_\Omega A\phi(t-\tau)w(\tau)\,dx\,d\tau$$

$$- \int_0^t \int_\Omega A\phi(t-\tau)Df(\tau)\,dx\,d\tau$$

$$= \int_0^t \int_\Omega \Delta\phi(t-\tau)w(\tau)\,dx\,d\tau$$

$$- \int_0^t \int_\Gamma f(\tau)\,(\gamma_1\phi(t-\tau))\,d\Gamma\,d\tau. \quad (6.11)$$

---

[1] The equation is the same as (6.3). We rewrite and give a special number to stress that the initial conditions belong to $\mathscr{D}(\Omega)$ and the affine term is zero.

Finally, the order of convolutions can be interchanged, so that

$$\int\limits_{\Omega} \int\limits_0^t \phi(t-\tau) \int\limits_0^{\tau} K(\tau-r)w(r)\,dr\,d\tau\,dx = \int\limits_{\Omega} \int\limits_0^t w(\tau) \int\limits_0^{t-\tau} K(t-\tau-r)\phi(r)\,dr\,d\tau\,dx.$$

We equate the left and the right sides. We get

$$\int\limits_{\Omega} \phi_0 w'(t)\,dx + \int\limits_{\Omega} \phi_1 w(t)\,dx = \int\limits_{\Omega} \phi(t)v_0\,dx + \int\limits_{\Omega} \phi'(t)u_0\,dx$$

$$- \int\limits_0^t \int\limits_{\Gamma} (\gamma_1\phi(t-r))\,f(r)\,d\Gamma\,dr$$

$$+ \int\limits_0^t \int\limits_{\Omega} \phi(t-r)F(r)\,dx\,dr. \qquad (6.12)$$

We recall that when the distribution is regular, the duality pairing is the integral of a product so that (6.12) coincides with (6.8), under the stringent conditions stated above on u_0, v_0, f, and F.

Now we keep ϕ_0 and ϕ_1 fixed in $\mathscr{D}(\Omega)$ and we write equality (6.12) for a sequence

$$u_n \to u_0 \in L^2(\Omega), \quad v_n \to v_0 \in H^{-1}(\Omega),$$
$$f_n \to f \in L^2(0,T;L^2(\Gamma)), \quad F_n \to F \in L^1(0,T;L^2(\Omega)).$$

Using continuous dependence on the initial data and on F and f (i.e., Theorem 2.4), property (6.7) and $\gamma_1\phi \in L^2(\Gamma \times (0,T))$, we can pass to the limit. The first integrals on both the sides converge, respectively, to

$$\langle\langle \phi_0, w'(t)\rangle\rangle, \quad \langle\langle \phi(t), v_0\rangle\rangle,$$

so that equality (6.8) holds for every $u_0 \in L^2(\Omega)$, $v_0 \in H^{-1}(\Omega)$, $f \in L^2(0,T;L^2(\Gamma))$ and $F \in L^1(0,T;L^2(\Omega))$. This proves the necessity part.

To prove sufficiency, we fix $u_0 \in L^2(\Omega)$, $v_0 \in H^{-1}(\Omega)$, $f \in L^2(0,T;L^2(\Gamma))$ and $F \in L^1((0,T),L^2(\Omega))$. Continuous dependence on the initial conditions (i.e., Theorem 2.4) and the direct inequality both applied to ϕ imply that the right-hand side of (6.8) is, for every $t \in [0,T]$, a linear continuous functional on $H_0^1(\Omega) \times L^2(\Omega)$. Hence, for every t there exists a *unique representation* of the right-hand side as

$$\langle\langle \phi_0(x), \tilde{\chi}(t)\rangle\rangle + \int\limits_{\Omega} \phi_1(x)\chi(x,t)\,dx \qquad (6.13)$$

with $\tilde{\chi}(t) \in H^{-1}(\Omega)$ and $\chi(t) \in L^2(\Omega)$. But, we have just proved that the special pair $(w(t), w'(t))$ satisfies the equality (6.8) and so it gives a representation of the form (6.13) of the right-hand side. *Unicity* of the representation implies that the pair $(\chi(t), \tilde{\chi}(t))$ must be equal to $(w(t), w'(t))$, hence $\chi(t)$ solves (6.1) and $\tilde{\chi}(t) = \chi'(t)$. $\square$

We sum up: now we have a test for the solutions, in which the set Ω appears explicitly.

Of course, Theorem 6.1 can be applied in particular when $f = 0$ and gives a test for solutions of (6.3) with "less regular" initial data, in $L^2(\Omega) \times H^{-1}(\Omega)$.

We shall be interested also in a formula similar to (6.8), with the solution ϕ of (6.9) replaced by $D_i\phi$ (D_i partial derivative). The function $D_i\phi$ is of class $C^\infty(\Omega \times (0, T))$ and solves the differential equation in (6.9) but in general $\gamma_1 D_i\phi$ will not be zero (see Remark 2.5). In spite of this, we have:

Lemma 6.2 *Let $F = 0$. Let ϕ solve (6.9) and ψ solve (6.3) with $\psi(0) = \psi_0 \in H_0^1(\Omega)$ and $\psi'(0) = \psi_1 \in L^2(\Omega)$. Let $\Omega_0 \subseteq \Omega$ be such that $\overline{\Omega}_0 \subseteq \Omega$. If $\psi(x, t) = 0$ in $\Omega \setminus \Omega_0$, for every $t \in [0, T]$ then the following equality holds, for every $t \in [0, T]$:*

$$\int_\Omega (D_i\phi_0)\, \psi'(t)\, dx + \int_\Omega (D_i\phi_1)\, \psi(t)\, dx = \int_\Omega (D_i\phi(t))\, \psi_1\, dx$$

$$+ \int_\Omega \left(D_i\phi'(t)\right) \psi_0\, dx. \tag{6.14}$$

The proof is a computation similar to the one above: first we insert a region Ω_I with C^∞ boundary

$$\overline{\Omega}_0 \subseteq \Omega_I \subseteq \overline{\Omega}_I \subseteq \Omega$$

so that the integrals on Ω are equal to those on Ω_I, since $\psi = 0$ on $\Omega \setminus \Omega_0$ and, as noted in Remark 2.5, $\gamma_1 D_i\phi$ is smooth on $\partial\Omega_I$ (in general different from zero).

The function ψ solves (6.3) on Ω_I with $\psi = 0$ on $\partial\Omega_I$ and we have also $\gamma_1\psi = 0$ on $\partial\Omega_I$.

We approximate ψ_0 and ψ_1 on Ω_I with functions $\psi_{0,n}$, $\psi_{1,n}$ in $\mathscr{D}(\Omega)$ and with support in Ω_I. We solve (6.3) on Ω_I with data $\psi_{0,n}$, $\psi_{1,n}$. The solution is denoted as ψ_n.

Note that $\psi_n = 0$ on $\partial\Omega$ so that the integration by parts in (6.10) and (6.11), with ϕ replaced by $D_i\phi$ on Ω_I gives the additional term (here γ_1 is the normal derivative on $\partial\Omega_I$)

$$\int_0^t \int_{\partial\Omega_I} (\gamma_0 D_i\phi(t - \tau))\, (\gamma_1 \psi_n(\tau))\, d(\partial\Omega_I)\, dt. \tag{6.15}$$

Now we pass to the limit on n, as in the proof of Theorem 6.1. The direct inequality on $\partial\Omega_I$ implies that $\gamma_1\psi_n \to \gamma_1\psi = 0$ and so both the integrals in (6.15) converge to zero and we get equality (6.14) on Ω_I, hence also on Ω.

6.2.1 Controllability and the Observation Inequality

In this section, we use the definition of the adjoint of an operator acting between Banach spaces. For clarity here, an index in the duality pairing, like $\langle\langle \cdot, \cdot \rangle\rangle_{H,H'}$, denotes the spaces. We confine ourselves to the case $L \in \mathcal{L}(H, K)$. The ADJOINT L^* belongs to $\mathcal{L}(K', H')$ and it is defined by the equality

$$\langle\langle Lh, k' \rangle\rangle_{K,K'} = \langle\langle h, L^*k' \rangle\rangle_{H,H'} \quad \forall h \in H, \ k' \in K'.$$

Continuity of L and L^* implies that it is sufficient to check the equality on dense subspaces of H and K'.

Theorem 1.2 holds also in Banach spaces. In particular, L has dense image when $\ker L^* = 0$ and it is surjective when L^* is COERCIVE, i.e., there exists $m_0 > 0$ such that $m_0|k'|_{K'} \leq |L^*k'|_{H'}$, for all $k' \in K'$.

The annihilator $R^\perp$ of a set $R \subseteq H$, where H is a Banach space, is the subspace of H' defined as follows:

$$R^\perp = \{h' \in H' : \langle\langle h, h' \rangle\rangle_{H,H'} = 0 \ \forall h \in R\}.$$

In particular we have:

- when R is a subspace of H, $R^\perp = 0$ if and only if R is dense in H.
- if $R = \mathrm{im}\,\Lambda$ then $R^\perp = \ker \Lambda^*$.

We use these properties in order to study controllability in a "dual" way. Formula (6.8) is the natural setting for this.

We consider (6.1) with $F = 0$ and null initial conditions. Let ϕ solve (6.9). We fix a time $T > 0$ at which we intend to study whether system (6.1) is controllable. Equality (6.8) takes the form:

$$\langle\langle \phi_0, w'(T) \rangle\rangle + \int_\Omega \phi_1 w(T)\,dx = -\int_0^T \int_\Gamma f(s)\gamma_1\phi(T-s)\,d\Gamma\,ds. \qquad (6.16)$$

Here ϕ_0 and ϕ_1 belong to $\mathcal{D}(\Omega)$.

This equality holds for every $f \in L^2(0, T; L^2(\Gamma))$. Now we keep f fixed, and we use ϕ_0 and ϕ_1 in $\mathcal{D}(\Omega)$ to approximate arbitrary elements, respectively, of $H_0^1(\Omega)$ and $L^2(\Omega)$. We get (6.16) for every solution of (6.3) with $F = 0$, $\phi_0 \in H_0^1(\Omega)$, and $\phi_1 \in L^2(\Omega)$. So: *If w solve (6.1) (with null initial conditions and affine term, then)*

(6.16) *holds for every* $f \in L^2(0, T; L^2(\Gamma))$, $\phi_0 \in H_0^1(\Omega)$ *and* $\phi_1 \in L^2(\Omega)$ *and the corresponding solution* ϕ *of* (6.3) *with* $F = 0$.

In order to interpret equality (6.16), let us introduce the real Banach spaces

$$H = L^2(0, T; L^2(\Gamma)), \quad K = H^{-1}(\Omega) \times L^2(\Omega)$$

and the operator Λ_T:

$$\Lambda_T f = (w'(T), w(T)) \ : \Lambda_T \in \mathscr{L}\left(L^2(0, T; L^2(\Gamma)), H^{-1}(\Omega) \times L^2(\Omega) \right).$$

Note that both H and K are Hilbert spaces, but the Hilbert space structure is not used since we don't identify K with its dual: we represent

$$K' = \left(H^{-1}(\Omega) \times L^2(\Omega) \right)' = H_0^1(\Omega) \times L^2(\Omega) \quad \text{while} \quad H' = L^2(0, T; L^2(\Gamma)).$$

We introduce the notation

$$\langle\langle \cdot, \cdot \rangle\rangle_1 = \langle\langle \cdot, \cdot \rangle\rangle_{K,K'} = \langle\langle \cdot, \cdot \rangle\rangle_{(H^{-1}(\Omega) \times L^2(\Omega)), (H_0^1(\Omega) \times L^2(\Omega))}$$

and the operator $(\Sigma_T(\phi_0, \phi_1))(t) = -\gamma_1 \phi(x, T - t)$:

$$\Sigma_T \in \mathscr{L}\left(H_0^1(\Omega) \times L^2(\Omega), L^2(0, T; L^2(\Gamma)) \right).$$

Then, equality (6.16) can be written as

$$\langle\langle \Lambda_T f, (\phi_0, \phi_1) \rangle\rangle_1 = -\int_0^T \int_\Gamma f(s)\gamma_1\phi(T - s)\,d\Gamma\,ds = \langle\langle f, \Sigma_T(\phi_0, \phi_1) \rangle\rangle_{H,H'}.$$

$$(6.17)$$

and this shows that

$$\Lambda_T^* = \Sigma_T, \quad \Sigma_T^* = \Lambda_T. \tag{6.18}$$

The reachable set at time T is

$$R_T = \operatorname{im} \Lambda_T :$$

approximate controllability at time T is the fact that im Λ_T is dense while we have exact controllability when Λ_T is surjective:

Theorem 6.2 *Let* $(\phi_0, \phi_1) \in H_0^1(\Omega) \times L^2(\Omega)$. *We have* $(\phi_0, \phi_1) \in (R_T)^\perp$ *if and only if the corresponding solution* ϕ *of* (6.3) *(with* $F = 0$) *has the additional property*

$$\gamma_1\phi = 0 \quad \forall t \in (0, T). \tag{6.19}$$

This result is often expressed as:

Corollary 6.1 *Approximate controllability is equivalent to* UNIQUE CONTINUATION *from the boundary which means that if a solution ϕ of Eq. (6.3) with $F = 0$, $\phi_0 \in H_0^1(\Omega)$, and $\phi_1 \in L^2(\Omega)$ satisfies the additional condition*

$$\gamma_1\phi = 0 \, on \, \Gamma \times (0, T)$$

then $\phi = 0$.

As to exact controllability, we use that the operator Λ_T is surjective if and only if its adjoint Σ_T is coercive:

Theorem 6.3 *The control process (6.1) (i.e., also (2.1)) is controllable at time T if and only if there exists $m_0 > 0$ such that*

$$m_0 \left(|\phi_0|^2_{H_0^1(\Omega)} + |\phi_1|^2_{L^2(\Omega)} \right) \leq \int_0^T \int_\Gamma |\gamma_1\phi(x, t)|^2 \, d\Gamma \, dt \qquad (6.20)$$

for every solution ϕ of (6.3) with $\phi_0 \in H_0^1(\Omega)$, $\phi_1 \in L^2(\Omega)$ (and $F = 0$).

Inequality (6.20) is called the OBSERVATION INEQUALITY or INVERSE INEQUALITY for the control process (6.1). A similar result in the case that the control is distributed in the domain is in [62].

Of course, Theorem 6.3 holds in particular if $K = 0$, i.e., for the memoryless wave equation.

6.3 The Observation Inequality for Systems with Memory

Informally, the goal of this section is the proof that the inverse inequality of the associated wave equation

$$\xi'' = \Delta\xi \, , \qquad \begin{cases} \xi(0) = \xi_0, & \xi'(0) = \xi_1, \\ \xi = 0 & \text{on the boundary} \end{cases} \qquad (6.21)$$

is inherited by the equation with persistent memory.

We introduce a second region Ω_1 with C^2 boundary and such that

$$\overline{\Omega} \subseteq \Omega_1.$$

We must be clear on the following facts, which will not be repeated:

- The control time of the associated wave equation (6.21) on Ω is denoted as T_0 and that on Ω_1 is denoted as T_1.
- system (6.1) will always be considered with null initial conditions and affine term;
- system (6.1) is always considered on Ω. So, for every T, even for $T = T_1$, $R_T \subseteq L^2(\Omega) \times H^{-1}(\Omega)$. Instead, the wave equation (6.21) will be considered either on Ω or on Ω_1, as specified;
- we use the restrictive condition

$$\Gamma = \partial\Omega , \quad w = f \text{ on } \Gamma = \partial\Omega ;$$

- γ_1 will be used to denote the trace of the exterior normal derivative on $\Gamma = \partial\Omega$ (the trace of the normal derivative on $\partial\Omega_1$ will not be needed);

Note that $T_1 > T_0$ but arbitrary close to it, by taking $\Omega_1 \setminus \Omega$ "small".
Either on the region Ω or on the region Ω_1 as explicitly specified, we consider

$$\phi'' = \Delta\phi + h\phi + \int_0^t K(t-r)\phi(r)\,dr \qquad \begin{cases} \phi(0) = \phi_0, \ \phi'(0) = \phi_1, \\ \phi = 0 \ \begin{cases} \text{either on } \partial\Omega \\ \text{or on } \partial\Omega_1. \end{cases} \end{cases} \qquad (6.22)$$

The result we are going to prove is:

Theorem 6.4 *The system with memory (6.1) (on the region Ω) is controllable in time T_1.*

Controllability of the wave equation *on the region Ω and at time T_0* implies that the reachable set R_T of (6.1) in $L^2(\Omega) \times H^{-1}(\Omega)$ is *closed with finite codimension for every $T \geq T_0$* (see Theorem 5.6). Hence, for the proof of Theorem 6.4 it is sufficient to prove $(R_{T_1})^\perp = 0$. A characterization of $(R_{T_1})^\perp$ is in Theorem 6.2: $(\phi_0, \phi_1) \in (R_{T_1})^\perp$ when $(\phi_0, \phi_1) \in H_0^1(\Omega) \times L^2(\Omega)$ and furthermore, the corresponding solution of (6.22) on Ω satisfies $\gamma_1\phi = 0$ in $L^2(\partial\Omega \times (0, T_1))$. Hence, the proof of Theorem 6.4 boils down to prove that *if $\phi(t)$ solves (6.22) on Ω then the following holds*:

$$\gamma_1\phi = 0 \text{ in } L^2(\partial\Omega \times (0, T_1)) \text{ implies } \phi(t) = 0 \text{ for } t \in [0, T_1].$$

The proof, from [52], is in three steps.

6.3.1 Step A: Extension and Derivatives of the Solutions

We prove two results which are natural by using the test of the solution in Theorem 6.1. To understand these theorems, we must bear in mind the following facts (see [51, Chap. 2]):

- let Ω have smooth boundary and let $\Omega_1 \supseteq \overline{\Omega}$. We extend with 0 to Ω_1 any $\phi \in H_0^1(\Omega)$: $\phi^e(x) = \phi(x)$ if $x \in \Omega$, $\phi^e(x) = 0$ if $x \in \Omega_1 \setminus \Omega$. Then $\phi^e \in H_0^1(\Omega_1)$;
- Let $D_i = \partial/\partial x_i$ be the i-th partial derivative (in the sense of distributions), i.e.,

$$\langle\langle \phi, D_i w \rangle\rangle = - \int_\Omega w\,(D_i \phi)\, dx. \tag{6.23}$$

We have: $D_i \in \mathcal{L}(H_0^1(\Omega), L^2(\Omega))$ and $D_i \in \mathcal{L}(L^2(\Omega), H^{-1}(\Omega))$.

On the region Ω, we consider the problem

$$\begin{cases} \psi'' = \Delta\psi + h\psi + \int\limits_0^t K(t - r)\psi(r)\,dr, \\ \psi(0) = \psi_0 \in H_0^1(\Omega), \ \psi'(0) = \psi_1 \in L^2(\Omega), \ \psi = 0 \text{ on } \partial\Omega. \end{cases} \tag{6.24}$$

The direct inequality implies that $\gamma_1 \psi \in L^2(0, T; L^2(\partial\Omega))$.

Let now $\Omega_1 \supseteq \overline{\Omega}$ be a region with C^2 boundary. We extend $\psi_0(x)$, $\psi_1(x)$, and $\psi(x, t)$ with 0 to Ω_1. The apex e denotes these extensions. We prove:

Theorem 6.5 *If the solution ψ of problem (6.24) on the region Ω has the additional property*

$$\gamma_1 \psi = 0 \text{ on } \Gamma = \partial\Omega$$

then the function $\psi^e = \psi^e(x, t)$ solves the following problem on the region Ω_1:

$$\psi^{e''} = \Delta\psi^e + h\psi^e + \int\limits_0^t K(t - r)\psi^e(r)\,dr, \quad \begin{cases} \psi^e(0) = \psi_0^e(x) \in H_0^1(\Omega_1), \\ (\psi^e)'(0) = \psi_1^e \in L^2(\Omega_1), \\ \psi^e = 0 \text{ on } \partial\Omega_1. \end{cases} \tag{6.25}$$

Proof It is obvious that

$$\psi^e(t) \in C([0, T]; H_0^1(\Omega_1)), \quad (\psi^e)'(t) = (\psi')^e(t) \in C([0, T], L^2(\Omega_1)).$$

So, $\psi^e(t)$ belongs to $C([0, T]; H_0^1(\Omega_1)) \cap C^1([0, T]; L^2(\Omega_1))$ and satisfies the initial and boundary conditions in (6.25). We must prove that $\psi^e(t)$ is a solution of problem (6.25) (on the region Ω_1).

In the following arguments, pay attention to the distinction between the regions Ω and Ω_1 and note that γ_1 always denotes normal derivative on the boundary of Ω.

There exist sequences $\{\psi_{0,n}\}$ and $\{\psi_{1,n}\}$ in $\mathscr{D}(\Omega_1)$ with the following properties: $\psi_{0,n}$ and $\psi_{1,n}$ have support in Ω and furthermore

$$(\psi_{0,n}, \psi_{1,n}) \to (\psi_0, \psi_1) \text{ in } H_0^1(\Omega) \times L^2(\Omega) \tag{6.26}$$

(of course, we have also $\psi_{0,n} \to \psi_0^e$ in $H_0^1(\Omega_1)$ and $\psi_{1,n} \to \psi_1^e$ on $L^2(\Omega_1)$).

Let $\psi_n(t)$ be the solution of (6.24) with data $(\psi_{0,n}, \psi_{1,n})$ on Ω (note: zero boundary condition on $\partial\Omega$).

Thanks to (6.26), we have $\psi_n(t) \to \psi(t)$ in $C([0, T]; H_0^1(\Omega)) \cap C^1([0, T]; L^2(\Omega))$ where $\psi(t)$ solves (6.24) on Ω. Hence, *thanks to the direct inequality,*

$$\gamma_1 \psi_n(t) \to \gamma_1 \psi = 0 \text{ in } L^2((\partial\Omega) \times (0, T)). \tag{6.27}$$

Let us fix *arbitrary* ϕ_0 and ϕ_1 in $\mathscr{D}(\Omega_1)$. We solve (6.25) on Ω_1 with these initial conditions and $\phi = 0$ on $\partial\Omega_1$. Let $\phi(t)$ be the solution. We know that $\phi(t)$ is a regular solution, in particular $\phi(t) \in H^2(\Omega_1)$, for every t.

Now we mimic the proof of Theorem 6.1: for every n, we multiply both the sides of (6.24) by $\phi(t - \tau)$ and we integrate on $\Omega \times (0, t)$ (note, on Ω and not Ω_1). We integrate by parts (in t the left side, in x the integral containing the laplacian) and we reorder the convolution. Using that $\psi_n(t) = 0$ on $\Gamma = \partial\Omega$ we get:

$$\int_0^t \int_\Gamma \phi(t - \tau) (\gamma_1 \psi_n(\tau)) \, d\Gamma \, d\tau$$

$$= \int_\Omega \left\{ \phi_0 \psi_n'(t) - \phi(t)\psi_{1,n} + \phi_1\psi_n(t) - \phi'(t)\psi_{0,n} \right\} dx$$

$$= \int_{\Omega_1} \left\{ \phi_0 \psi^{e'}{}_n(t) - \phi(t)\psi_{1,n}^e + \phi_1\psi_n^e(t) - \phi'(t)\psi_{0,n}^e \right\} dx.$$

The last equality holds since the extensions to $\Omega_1 \setminus \Omega$ are the extensions with zero.

Now we pass to the limit for $n \to +\infty$. The integral on $\Gamma = \partial\Omega$ tends to zero (see 6.27)) and we remain with the equality

$$\int_{\Omega_1} \phi_0 \psi^{e'}(t) \, dx + \int_{\Omega_1} \phi_1 \psi^e(t) \, dx = \int_{\Omega_1} \phi(t)\psi_1^e \, dx + \int_{\Omega_1} \phi'(t)\psi_0^e \, dx. \tag{6.28}$$

Thanks to the fact that $\psi_0^e \in H_0^1(\Omega_1)$, $\psi_1^e \in L^2(\Omega_1)$, the duality pairings in (6.8) are integrals of products. Hence, ψ^e passes the text of Theorem 6.1 for a solution of (6.25) *on the larger region Ω_1, and homogeneous boundary conditions on $\partial\Omega_1$.* This ends the proof.

Theorem 6.6 *Let $\Gamma = \partial\Omega$ and let ψ solve Eq. (6.24) on $[0, T]$, with initial conditions $(\psi_0, \psi_1) \in H_0^1(\Omega) \times L^2(\Omega)$. We assume the existence of a region Ω_0 such that*

$$\overline{\Omega}_0 \subseteq \Omega \text{ and } \psi(x, t) = 0 \text{ on } \Omega \setminus \Omega_0.$$

Then $D_i\psi \in C([0, T]; L^2(\Omega)) \cap C^1([0, T]; H^{-1}(\Omega))$ solves Eq. (6.24) on $[0, T]$, with the conditions

$$D_i\psi(x, 0) = D_i\psi_0(x) \in L^2(\Omega),$$
$$D_i\psi'(x, 0) = D_i\psi_1(x) \in H^{-1}(\Omega),$$
$$(D_i\psi) = 0 \text{ on } \partial\Omega.$$

Proof The function ψ satisfies the condition of Lemma 6.2, so that (6.14) holds. The function $\phi(x, t)$ is of class C^∞ for every t and $\psi(x, t)$ and $\psi'(x, t)$ have compact support for every t. So, using (1.24), equality (6.14) can be rewritten as:

$$\langle\langle\phi_0(x), D_i\psi'(x, t)\rangle\rangle + \int_\Omega \phi_1(x)D_i\psi(x, t)\,dx$$

$$= \langle\langle\phi(x, t), D_i\psi_1(x)\rangle\rangle + \int_\Omega \phi'(x, t)D_i\psi_0(x)\,dx.$$

The functions ϕ_0 and ϕ_1 are arbitrary in $\mathscr{D}(\Omega)$, so that $D_i\psi$ passes the test of Theorem 6.1 for a solution of (6.25) on Ω, with zero boundary condition and initial data $D_i\psi_0 \in L^2(\Omega)$, $D_i\psi_1 \in H^{-1}(\Omega)$.

6.3.2 Step B: Propagation of Singularities

We assume that the initial conditions of (6.22) satisfy $\phi_0 \in L^2(\Omega)$ and $\phi_1 \in H^{-1}(\Omega)$ so that

$$\phi(t) \in C(0, T; L^2(\Omega)) \cap C^1(0, T; H^{-1}(\Omega)).$$

Let us assume furthermore that the initial conditions ϕ_0 and ϕ_1 have compact support in Ω. Due to the finite propagation speed, $\phi(x, t)$ and $\phi'(x, t)$ will have compact support in Ω at least if t is small. Hence, there exists a region Ω_0 such that

$$\overline{\Omega}_0 \subseteq \Omega \text{ and } \phi(t) = 0 \text{ in } \Omega \setminus \Omega_0, \tag{6.29}$$

at least for small t. We wonder how long this can last. We prove the following result, which will be improved later on:

Theorem 6.7 *Let $\overline{\Omega}_0 \subseteq \Omega$. Let $\phi_0 \in L^2(\Omega)$, $\phi_1 \in H^{-1}(\Omega)$ have compact support in Ω_0. Let $\phi(t)$ be the corresponding solution of (6.22).*

Assume that the observation inequality holds at time T for the wave equation on Ω and that $\phi(t) = 0$ on $\Omega \setminus \Omega_0$, for $t \in [0, T]$. Then the initial conditions ϕ_0 and ϕ_1 (which are zero in $\Omega \setminus \Omega_0$) have the following additional regularity:

$$\phi_0 \in H_0^1(\Omega), \quad \phi_1 \in L^2(\Omega)$$

so that we have also $\phi \in C([0, T]; H_0^1(\Omega)) \cap C^1([0, T]; L^2(\Omega))$. Furthermore, there exists a constant $K = K_T$ such that

$$\left(|\phi_0|^2_{H_0^1(\Omega)} + |\phi_1|^2_{L^2(\Omega)}\right) \leq K |\phi|^2_{C(0,T;L^2(\Omega))}. \tag{6.30}$$

Proof Let $\rho(x)$ be a positive C^∞ function with compact support in $|x| < 1$ and such that

$$\int_{\mathbf{R}^n} \rho(x)\,dx = 1.$$

Let

$$\phi_\varepsilon(x, t) = \int_{\mathbf{R}^n} \rho\left(\frac{y}{\varepsilon}\right) \phi(x - y, t)\,dy = \int_{\mathbf{R}^n} \rho\left(\frac{x - y}{\varepsilon}\right) \phi(y, t)\,dy$$

(in these integrals, we put $\phi(x, t) = 0$ if $x \in \mathbf{R}^n \setminus \Omega$).

We list the following known properties:

- There exists $\varepsilon_0 > 0$ such that for $\varepsilon \in (0, \varepsilon_0)$ the support of $x \mapsto \phi_\varepsilon(x, t)$ is contained in a region Ω_I, for every $t \in [0, T]$, and Ω_I satisfies

$$\overline{\Omega}_0 \subseteq \Omega_I \subseteq \overline{\Omega}_I \subseteq \Omega.$$

We stress the fact that Ω_I does not depend on ε (for $0 < \varepsilon < \varepsilon_0$).
- $\phi_\varepsilon(t)$ and $\phi_\varepsilon'(t)$ belong to $\mathcal{D}(\Omega)$, for every $t \geq 0$.
- $\lim_{\varepsilon \to 0^+} \phi_\varepsilon = \phi$ in $C([0, T]; L^2(\Omega)) \cap C^1([0, T]; H^{-1}(\Omega))$. So (for a suitable M):

$$\begin{cases} |\phi_\varepsilon|_{C([0,T];L^2(\Omega))} \leq M|\phi|_{C([0,T];L^2(\Omega))} \\ \lim_{\varepsilon \to 0^+}(\phi_\varepsilon(0), \phi_\varepsilon'(0)) = (\phi_0, \phi_1) \text{ in } L^2(\Omega) \times H^{-1}(\Omega) \end{cases}. \tag{6.31}$$

- ϕ_ε is a regular solution of (6.22) on Ω.

In order to prove the theorem, we show[2]

$$w - \lim_{\varepsilon \to 0+} \left(\phi_\varepsilon(0), \phi'_\varepsilon(0) \right) \text{ exists in } H^1_0(\Omega) \times L^2(\Omega).$$

Then, we have the required property $(\phi_0, \phi_1) \in H^1_0(\Omega) \times L^2(\Omega)$ from the second statement of (6.31).

We use the property that ϕ_ε is a regular solution of (6.22) on Ω and we represent ϕ_ε on Ω as the sum of two functions:

$$\phi_\varepsilon = \psi_\varepsilon + \zeta_\varepsilon$$

where ψ_ε and ζ_ε solve, respectively:

$$\psi''_\varepsilon = \Delta \psi_\varepsilon + h\phi_\varepsilon + \int_0^t K(t-s)\phi_\varepsilon(s)\,ds, \quad \begin{cases} \psi_\varepsilon(t) = 0, \quad \psi'_\varepsilon(0) = 0, \\ \psi_\varepsilon = 0 \text{ on } \Gamma = \partial\Omega, \end{cases} \quad (6.32)$$

$$\zeta''_\varepsilon = \Delta \zeta_\varepsilon, \quad \begin{cases} \zeta_\varepsilon(t) = \phi_\varepsilon(0), \quad \zeta'_\varepsilon(0) = \phi'_\varepsilon(0), \\ \zeta_\varepsilon = 0 \text{ on } \Gamma = \partial\Omega. \end{cases} \quad (6.33)$$

Note that (6.32) is a wave equation with the affine term which depends on ϕ_ε.

Using the inequality in (6.31) and the direct inequality of the *wave equation* applied to (6.32) we get (the constants M are not the same)

$$|\gamma_1 \psi_\varepsilon|^2_{L^2(0,T;L^2(\Gamma))} \leq M|h\phi_\varepsilon + K * \phi_\varepsilon|^2_{L^1(0,T;L^2(\Omega))} < M|\phi|^2_{C([0,T];L^2(\Omega))}. \quad (6.34)$$

The *inverse inequality* of the *wave equation* applied to (6.33) gives (here $m > 0$)

$$m \left(|\phi_\varepsilon(0)|^2_{H^1_0(\Omega)} + |\phi'_\varepsilon(0)|^2_{L^2(\Omega)} \right) \leq |\gamma_1 \zeta_\varepsilon|^2_{L^2(0,T;L^2(\Gamma))}. \quad (6.35)$$

The fact $\phi_\varepsilon = 0$ in $\Omega \setminus \Omega_I$ shows that

$$\psi_\varepsilon(x,t) = -\zeta_\varepsilon(x,t) \text{ in } \Omega \setminus \Omega_I \text{ and so } \gamma_1 \psi_\varepsilon = -\gamma_1 \zeta_\varepsilon \text{ in } L^2(0,T;L^2(\partial\Omega)). \quad (6.36)$$

Using (6.36), we can combine the inequalities (6.34) and (6.35) to get

$$\left(|\phi_\varepsilon(0)|^2_{H^1_0(\Omega)} + |\phi'_\varepsilon(0)|^2_{L^2(\Omega)} \right) \leq \frac{M}{m} |\phi|^2_{C(0,T;L^2(\Omega))}. \quad (6.37)$$

So, the sets $\{\phi_\varepsilon(x,0)\}$ and $\{\phi'_\varepsilon(x,0)\}$ are bounded in $H^1_0(\Omega)$ and $L^2(\Omega)$ and this implies (for a suitable sequence $\varepsilon_n \to 0+$) weak convergence, respectively, in $H^1_0(\Omega)$ and in $L^2(\Omega)$. As we already know convergence to ϕ_0 and ϕ_1 we get

---

[2] in fact for a suitable $\varepsilon_n \to 0+$ and this is sufficient for the proof, but it is also possible to deduce that the limit for $\varepsilon \to 0+$ exists.

$\phi_0 \in H_0^1(\Omega)$ and $\phi_1 \in L^2(\Omega)$. Inequality (6.30) follows passing to the limit for $\varepsilon \to 0+$ in (6.37) and using weak lower semicontinuity of the norms (see (1.17)).

This gives the result.

The previous result can be interpreted as saying that any "singularity" of the initial conditions in Ω_0 sooner or later has to hit the boundary of Ω, in fact in a time shorter then the control time of the associated wave equation.

6.3.3 Step C: End of the Proof

We recall the apex ^{e} to denote extension with 0 to Ω_1 of elements of $H_0^1(\Omega)$ and $L^2(\Omega)$ and we consider the following subspace $\mathcal{S} \subseteq C([0, T]; H_0^1(\Omega_1)) \cap C^1([0, T]; L^2(\Omega_1))$:

$$\mathcal{S} = \left\{ \phi^e(t) : \left(\phi_{|\Omega}^e(0), \left(\phi_{|\Omega}^e \right)'(0) \right) = (\phi_0, \phi_1) \in H_0^1(\Omega) \times L^2(\Omega), \ (\phi_0, \phi_1) \perp R_{T_1} \right\}$$

(we intend that $\phi(t) = \phi_{|\Omega}^e(t)$ solves (6.22) on Ω) so that

$$\dim \mathcal{S} = \dim \left(R_{T_1} \right)^{\perp} < +\infty.$$

We prove:

Lemma 6.3 *The subspace $\mathcal{S}$ is invariant under any partial derivative D_i.*

Proof We fix any $(\phi_0, \phi_1) \perp R_{T_1}$ and we consider the corresponding solution $\phi(t) \in C([0, T_1]H_0^1(\Omega)) \cap C^1([0, T_1]; L^2(\Omega))$ of (6.22). Thanks to the condition $\gamma_1 \phi(t) = 0$ in $L^2(\Gamma \times (0, T_1))$, we conclude from Theorem 6.5 that $\phi^e(t)$ solves Eq. (6.22) on Ω_1, with initial condition $(\phi_0^e, \phi_1^e) \in H_0^1(\Omega_1) \times L^2(\Omega_1)$. This solution is zero in $\Omega_1 \setminus \Omega$ so that, from Theorem 6.6, $D_i \phi^e(t)$ solves Eq. (6.22) on Ω_1 and initial conditions $(D_i \phi_0^e, D_i \phi_1^e) \in L^2(\Omega_1) \times H^{-1}(\Omega_1)$.

Now we use that $D_i \phi^e(t)$ is zero on $\Omega_1 \setminus \Omega$ for a time *as long as the control time of the wave equation on Ω_1.* So, $(D_i \phi_0^e, D_i \phi_1^e) \in H_0^1(\Omega_1) \times L^2(\Omega_1)$.

This implies that if $(\phi_0, \phi_1) \perp R_{T_1}$ then $\phi_0^e \in H^2(\Omega_1)$, $\phi_1^e \in H^1(\Omega_1)$.

The previous procedure can be iterated: let us insert Ω_I between Ω and Ω_1: $\overline{\Omega} \subseteq \Omega_I \subseteq \overline{\Omega}_I \subseteq \Omega_1$ and

$$\psi^e(t) = D_i \phi^e(t) \ \text{ so that } \ \begin{cases} \psi^e(0) = \psi_0^e = D_i \phi_0^e \in H_0^1(\Omega_1), \\ (\psi^e)'(0) = \psi_1^e = D_i \phi_1^e \in L^2(\Omega). \end{cases}$$

We consider $D_j \psi^e(t)$ which solves Eq. (6.22) on Ω_1 and it is zero on $(\Omega_1 \setminus \Omega_I) \times [0, T_1]$. We conclude that $D_j \psi_0^e = D_j D_i \phi_0^e \in H_0^1(\Omega_1)$, for every i and j, i.e., $\phi_0^e \in H^3(\Omega_1)$ and analogously $\phi_1^e \in H^2(\Omega_1)$.

The procedure can be iterated and we conclude ϕ_0^e and ϕ_1^e belong to $H^k(\Omega_1)$, for every k, i.e., they are (represented by a) $C^\infty(\Omega_1)$-function, and are zero on $\Omega_1 \setminus \Omega$.

In particular, $\phi^e(t) \in C^\infty(\Omega_1 \times [0, +\infty))$ (see Remark 2.5) and for every i we have $(D_i \phi^e(t)) = 0$ on $\partial \Omega$ and so $\gamma_1 (D_i \phi(t)) = 0$.

We recapitulate: the elements of $R_{T_1}^\perp$ are of class C^∞ and $(D_i \phi_0, D_i \phi_1) \perp R_{T_1}$ so that D_i transforms $\mathscr{S}$ into itself.

Now we prove that $\mathscr{S} = 0$ so that we have also $(R_{T_1})^\perp = 0$, as we wanted to prove.

Let us fix one of the partial derivatives, for example D_1. As it is a linear transformation from the *finite dimensional* linear space $\mathscr{S}$ to itself, we have either $\mathscr{S} = 0$ or the transformation must have at least one eigenvalue, i.e., there must exist λ and $\phi \in \mathscr{S}$ such that

$$\phi \neq 0, \quad D_1 \phi = \lambda \phi. \tag{6.38}$$

We prove that any solution of the differential equation in (6.38) is zero, so that $\mathscr{S} = 0$ and $R_{T_1} = L^2(\Omega) \times H^{-1}(\Omega)$. We introduce the notations

$$y = x_1, \quad z = (x_2, \dots, x_n, t).$$

In Fig. 6.1, the vertical axis is the axis of the variable $y = x_1$, while the variable z (which includes the time) is represented on the horizontal axis (note that the variable $y = x_1$ is listed first in (y, z)).

Let us fix any $(y_0, z_0) \in \Omega \times (0, T)$. We prove that $\phi = 0$ in a "small cylinder" of the space (y, z), where z varies in a neighborhood N of z_0 and $y \in (y_3, y_1)$ with $y_3 < y_0 < y_1$.

Due to the fact that $\overline{\Omega} \subseteq \Omega_1$, we can find y_1 such that $(y_1, z_0) \in \Omega_1 \setminus \overline{\Omega}$ as in Fig. 6.1, left, and we construct a small cylinder as in Fig. 6.1, right, whose upper part, say from the level y_2 to y_1, is contained in $\Omega_1 \setminus \overline{\Omega}$ and $y_3 < y_0$.

The neighborhood N is the intersection of the cylinder with the hyperplane $y = y_0$.

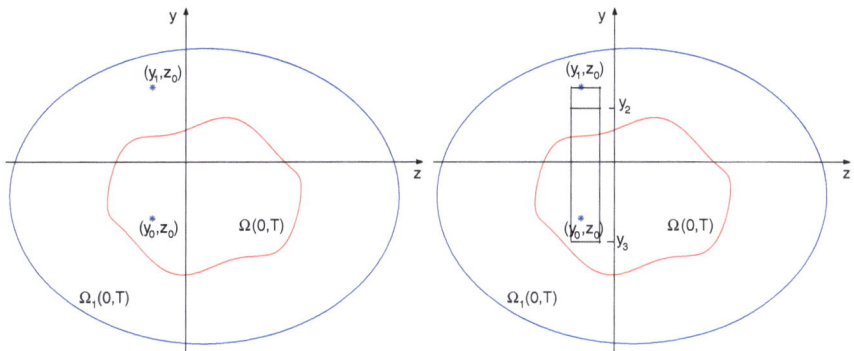

Fig. 6.1 The construction of the cylinder

Now choose any test function $\eta = \eta(z)$, independent of y, of class C^∞ and compact support contained in N and we consider the function

$$y \mapsto \int_N \phi(y, z)\eta(z)\,dz = \zeta(y)$$

We get a function defined on a segment which contains (y_3, y_1). This segment contains y_0.

Equality (6.38) implies that the function $\zeta(y)$ of the scalar variable y is differentiable and solves the ordinary differential equation

$$\zeta' = \lambda\zeta.$$

The fact that $\zeta(y) = 0$ for $y \in [y_2, y_1]$ proves that $\zeta(y) = 0$ everywhere, in particular for $y = y_0$, and this for every test function η with support in N. So we have $\phi(y_0, z) = 0$ in N and, moving y_0 "up and down", $\phi(y, z) = 0$ in the "cylinder". So, $\phi = 0$ everywhere in Ω, since (y_0, z_0) is arbitrary in Ω.

In conclusion, we proved Theorem 6.4 and so, combining with Theorem 6.3, we proved also the inverse inequality for the system with memory:

Theorem 6.8 *Let $\Gamma = \partial\Omega$. There exists a time T_1 such that system (6.1) is controllable. The time T_1 is the control time of the associated wave equation in any region Ω_1 which contains $\overline{\Omega}$.*

For every $T \geq T_1$ there exist $m > 0$ and M such that

$$m\left(|\phi_0|^2_{H_0^1(\Omega)} + |\phi_1|^2_{L^2(\Omega)}\right) \leq \int_{\Gamma\times[0,T]} |\gamma_1\phi|^2\,d\Gamma\,dt \leq M\left(|\phi_0|^2_{H_0^1(\Omega)} + |\phi_1|^2_{L^2(\Omega)}\right).$$

Remark 6.1 The paper [52] has a proof of the fact that R_{T_1} is closed with finite codimension which do not use moment methods (see also [74]). A third different proof is in [74]. The proof in [52] has $K = K(t, s)$. The case $K(t, s) = K(t - s)$ is by far the most important in applications, but also the case $P(t)K(t - s)$ is important, since it corresponds to bodies which, being part of a chain of mechanisms, are subject to external tractions. See [79] for a moment approach to this case.

6.4 Problems[3] to Chap. 6

6.1. Let w and ϕ solve $u'' = u_{xx}$ for $x > 0$, respectively with initial conditions $w(x, 0) = 0$, $w'(x, 0) = 0$, $\phi(x, 0) = \phi_0(x) \in \mathscr{D}(0, +\infty)$, $\phi'(x, 0) = \phi_1(x) \in \mathscr{D}(0, +\infty)$, and boundary conditions $w(0, t) = f(t) \in H^2(0, +\infty)$, $\phi(0, t) = 0$. Justify the integration by parts of the left-hand side of the equality

---

[3] Solutions at the address http://calvino.polito.it/~lucipan/materiale_html/P-S-CH-6.

$$\int_0^T \int_0^{+\infty} \phi(x, T - s) \left[w''(x, s) - w_{xx}(x, s)\right] dx \, ds = 0$$

and show that

$$\int_0^{+\infty} \left[\phi_0(x)w'(x, T) + \phi_1(x)w(x, T)\right] dx = + \int_0^{+\infty} \phi_x(0, T - s)f(s) \, ds.$$

Compare with Problem 4.2 and with formula (6.8) and explain the sign of the right-hand side.

6.2. Use (2.43) and Picard iteration to prove the direct inequality for system (2.43).

6.3. In the sufficiency part of the proof of Theorem 6.1, the function $t \mapsto (\chi(t), \tilde{\chi}(t))$ has been defined. This function takes values in $L^2(\Omega) \times H^{-1}(\Omega)$ (see formula (6.13)). Use the fact that the right-hand side of (6.8) is a continuous function of time and prove that $(\chi(t), \tilde{\chi}(t)) \in C([0, T]; L^2(\Omega) \times H^{-1}(\Omega))$.

6.4. This and the next problem show that the inverse inequality is an "observability property". Let us consider the problem

$$\phi'' = \phi_{xx}, \quad x \in (0, 1), \quad \begin{cases} \phi(x, 0) = \phi_0(x) \in C^1(0, 1), \; \phi'(x, 0) = 0 \\ \phi(0, t) = \phi(1, t) = 0. \end{cases} \tag{6.39}$$

Prove the existence of initial conditions $\phi_0 \neq 0$ such that $\phi_x(1, t) \equiv 0$ for $t \in [0, T]$, provided that $T < 1$ and prove that if $\phi_x(1, t) \equiv 0$ for $t \in [0, 1]$ then $\phi_0 = 0$.

6.5. Do the same as in Problem 6.4 when the initial conditions are $\phi(x, 0) = 0$, $\phi'(x, 0) = \phi_1(x) \in C(0, 1)$.

6.6. Use the results in Problem 2.4. Show that the inverse inequality does not hold for the wave equation if $\Omega = (0, \pi) \times (0, \pi)$ and $\Gamma = (0, 1) \times \{0\}$.

6.7. Let ϕ_0 and ϕ_1 and T be as in Theorem 6.7. Prove the existence of M such that the following holds, for every $t \in [0, T]$:

$$|\phi'(t)|_{L^2(\Omega)} \leq M |\phi|_{C([0,T];L^2(\Omega))}.$$

6.8. Let $\phi_0 \in H_0^1(\Omega)$, $\phi_1 \in L^2(\Omega)$. Use the Volterra integral equation (2.43) to represent the solutions of (6.3). For simplicity let $c_0 = 1$ and $h = 0$. Choose any $T > 0$. Consider the set $\mathcal{K}$ of the solutions such that $|\phi_0|_{H_0^1(\Omega)} \leq 1$, $|\phi_1|_{L^2(\Omega)} \leq 1$. Prove that $\mathcal{K}$ is bounded in $C([0, T]; L^2(\Omega))$ and equicontinuous.

6.9. In the space $C([0, T]; L^2(\Omega))$ the set $\mathcal{K}$ in Problem 6.8 is both bounded and equicontinuous. It is known that boundedness and equicontinuity of a subset $C([0, T]; H)$ implies compactness if $\dim H < +\infty$. Prove that this is not true if dim $H = +\infty$ (in spite of this negative result the special set $\mathscr{K}$ in Problem 6.8 is relatively compact, i.e., any sequence has convergent subsequences in $C([0, T]; L^2(\Omega))$), since its elements take value in $H^1(\Omega)$, which is compactly embedded in $L^2(\Omega)$, see [23, p. 266]).

6.10. Both for the system with memory and for the associated wave equation, study the following problem. Let controllability holds at a certain time T with the control acting on $\Gamma \subseteq \partial\Omega$. Let $\mathscr{H}_T$ be the subset of $(\phi_0, \phi_1) \in H_0^1(\Omega) \times L^2(\Omega)$ such that the corresponding solution of (2.43) satisfies (M is a fixed number)

$$\int_0^T \int_\Gamma |\gamma_1 \phi|^2 \, dG_T \leq M.$$

Prove that $\mathscr{H}_T$ is closed and decide whether it is also bounded.
    Discuss the role of the assumption of controllability in this problem.

6.11. The derivation of the inverse inequality for the wave equation uses conservation of energy. So, Theorem 6.8 might suggest conservation of energy for the system with memory.
    Examine the system

$$\phi' = \int_0^t e^{\alpha(t-s)} \Delta\phi(s) \, ds \qquad\qquad (6.40)$$

(with zero boundary conditions) and show that the quantity

$$|\phi'(t)|^2 + \int_\Omega |\nabla\phi|^2 \, dx$$

is not constant along the solutions.

References

1. Aubin, J.-P.: Applied Functional Analysis. Pure and Applied Mathematics, 2nd edn. Wiley, New York (2000)
2. Avdonin, S.A., Ivanov, S.A.: Families of Exponentials. The Method of Moments in Controllability Problems for Distributed Parameter Systems. Cambridge University Press, New York (1995)
3. Avdonin, S., Pandolfi, L.: Temperature and heat flux dependence/independence for heat equations with memory. In: Sipahi, R., Vyhlidal, T., Pepe, P., Niculescu, S.-I. (eds.) Time Delay Systems—Methods, Applications and New Trend. Lecture Notes in Control and Inform. Sci, 423, pp. 87–101. Springer, New York (2012)
4. Avdonin, S., Pandolfi, L.: Simultaneous temperature and flux controllability for heat equations with memory. Q. Appl. Math. **71**, 339–368 (2013)
5. Balachandran, K.: Controllability of functional semilinear integrodifferential systems in Banach spaces. J. Mat. Anal. Appl. **255**, 447–457 (2001)
6. Balachandran, K., Sakthivel, R.: Controllability of functional semilinear integrodifferential systems in Banach spaces. J. Math. Anal. Appl. **255**, 447–457 (2001)
7. Balakrishnan, A.V.: Applied Functional Analysis. Applications of Mathematics, vol. 3, 2nd edn. Springer, New York (1981)
8. Barbu, V., Iannelli, M.: Controllability of the heat equation with memory. Diff. Integr. Equat. **13**, 1393–1412 (2000)
9. Bardos, C., Lebeau, G., Rauch, J.: Sharp sufficient conditions for observation, control and stabilization of waves from the boundary. SIAM J. Control Optim. **305**, 1024–1065 (1992)
10. Belleni-Morante, A.: An integro-differential equation arising from the theory of heat conduction in rigid materials with memory. Boll. Un. Mat. Ital. **15–B**(5), 470–482 (1978)
11. Brezis, H.: Functional Analysis Sobolev Spaces and Partial Differential Equations. Universiotext, Springer, New York (2011)
12. Burq, N.: Contrôlabilité exacte des ondes dans des ouvertes peu réguliers. Asympt. Anal. **14**, 157–191 (1997)
13. Cattaneo, C.: Sulla conduzione del calore. Atti Semin. Mat. e Fis. Univ. di Modena **3**, 83–101 (1948)
14. Cazenave, T.: On the propagation of confined waves along geodesics. J. Mat. Anal. Appl. **146**, 591–603 (1990)
15. Christensen, O.: Frames and Bases. An Introductory Course. Birkhäuser Boston, MA (2008)
16. Corduneanu, C.: Integral Equations and Applications. Cambridge U.P, Cambridge (1991)

© The Author(s) 2014

L. Pandolfi, *Distributed Systems with Persistent Memory*, SpringerBriefs in Control, Automation and Robotics, DOI 10.1007/978-3-319-12247-2

17. Curtain, R.F., Zwart, H.J.: An Introduction to Infinite Dimensional Linear Systems Theory. Texts in Applied Mathematics, vol. 21. Springer, New York (1995)
18. Dafermos, C.M.: Asymptotic stability in viscoelasticity. Arch. Rational Mech. Anal. **37**, 297–308 (1970)
19. Dafermos, C.M.: Contraction semigroups and trend to equilibrium in continuum mechanics. In: Applications of Methods of Functional Analysis to Problems in Mechanics, Lecture Notes in Mathematics 503, pp. 295–306. Springer, Berlin (1976)
20. Dautray, R., Lions, J.-L.: Mathematical Analysis and Numerical Methods for Science and Technology. Physical Origins and Classical Methods, vol. 1. Springer, Berlin (1990)
21. De Kee, D., Liu, Q., Hinestroza, J.: Viscoelastic (non-fickian) diffusion. Can. J. Chem. Eng. **83**, 913–929 (2005)
22. Desch, W., Grimmer, R.C.: Initial-boundary value problems for integro-differential equations. J. Integr. Equat. **10**, 73–97 (1985)
23. Dunford, N., Schwartz, J.T.: Linear operators part I: general theory. Wiley Classic Library Edition, Wiley, New York (1988)
24. Ervedoza, S., Zuazua, E.: Numerical Approximation of Exact Controls for Waves. Springer Briefs in Mathematics. Springer, New York (2013)
25. Fabrizio, M., Giorgi, C., Pata, V.: A new approach to equations with memory. Arch. Rational Mech. Anal. **198**, 189–232 (2010)
26. Fardigola, D.: Controllability problems for the $1 - d$ wave equation on a half axis with the Dirichlet boundary condition. ESAIM Control Optim. Calc. Var. **18**, 748–773 (2012)
27. Fattorini, H.O.: Second Order Linear Differential Equations in Banach Spaces. North-Holland, Amsterdam (1985)
28. Fisher, G.M.C., Gurtin, M.E.: Wave propagation in the linear theory of viscoelasticity. Quart. Appl. Math. **23**, 257–263 (1965)
29. Fu, X., Yong, J., Zhang, X.: Controllability and observability of the heat equation with hyperbolic memory kernel. J. Diff. Equat. **247**, 2395–2439 (2009)
30. Fuhrmann, P.A.: On weak and strong reachability and controllability of infinite dimensional linear systems. J. Optim. Theory Appl. **9**, 77–89 (1972)
31. Fujishiro, K., Yamamoto, M.: Approximate controllability for fractional diffusion equations by interior control. Appl. Anal. **93**, 1793–1810 (2014)
32. Fujiwara, D.: Concrete characterization of the domains of fractional powers of some elliptic differential operators of the second order. Proc. Japan Acad. **43**, 82–86 (1967)
33. Giorgi, G., Gentili, G.: Thermodynamic properties and stability for the heat flux equation with linear memory. Quart. Appl. Math. **51**, 343–362 (1993)
34. Gohberg, I.C., Krejn, M.G.: Opèrateurs Linèaires non Auto-adjoints dans un Espace Hilbertien. Dunod, Paris (1971)
35. Goldberg, S.: Unbounded Linear Operators. Theory and Applications. McGraw-Hill Book Co., New York (1966)
36. Gubreev, G.M., Volkova, M.G.: One remark about the unconditional exponential bases and cosine bases connected with them. Methods Funct. Anal. Topol. **14**, 330–333 (2008)
37. Guerrero, S., Imanuvilov, O.Y.: Remarks on non controllability of the heat equation with memory. ESAIM Control Optim. Calc. Var. **19**, 288–300 (2013)
38. Guo, B.-Z., Zhang, G.-D.: On spectrum and Riesz basis property for one-dimensional wave equation with Boltzmann damping. ESAIM Control Optim. Calc. Var. **18**, 889–913 (2012)
39. Gurtin, M.E., Pipkin, A.G.: A general theory of heat conduction with finite wave speed. Arch. Ration. Mech. Anal. **31**, 113–126 (1968)
40. Haak, B.H., Jacob, B., Partington, J.R., Pott, S.: Admissibility and controllability of Diagonal Volterra equations with scalar inputs. J. Diff. Equat. **246**, 4423–4440 (2009)
41. Halanay, A., Pandolfi, L.: Lack of controllability of the heat equation with memory. Syst. Control Lett. **61**, 999–1002 (2012)
42. Halanay, A., Pandolfi, L.: Lack of controllability of thermal systems with memory. Evol. Equ. Control Theory. **3**, 485–497 (2014)

43. Halanay, A., Pandolfi, L.: Approximate controllability and lack of controllability to zero of the heat equation with memory, submitted, arXiv:1404.2745
44. Hassel, A., Tao, T.: Erratum for upper and lower bounds for normal derivatives of Dirichlet eigenfunctions. Math. Res. Lett. **17**, 793–794 (2010)
45. Heil, C.: A Basis Theory Primer. Birkhäuser, Springer, New York (2011)
46. Herrera, I., Gurtin, M.E.: A correspondence principle for viscoelastic wave propagation. Quart. Appl. Math. **22**, 360–364 (1965)
47. Ivanov, S., Pandolfi, L.: Heat equations with memory: lack of controllability to rest. J. Mat. Anal. Appl. **355**, 1–11 (2009)
48. Jacob, B., Partington, R.: On controllability of diagonal systems with one-dimensional input space. Syst. Control Lett. **55**, 321–328 (2006)
49. Ji, S.: Approximate controllability of semilinear nonlocal fractional differential systems via an approximating method. Appl. Math. Comput. **236**, 43–53 (2014)
50. Kalman, R.E., Falb, P.L., Arbib, M.A.: Topics in Mathematical Systems Theory. McGraw-Hill book Co., New York (1969)
51. Kesavan, S.: Topics in Functional Analysis and Applications. Wiley, New York (1989)
52. Kim, J.U.: Control of a second-order integro-differential equation. SIAM J. Control Optim. **31**, 101–110 (1993)
53. Komornik, V.: Exact Controllability and Stabilization. The Multiplier Method. RAM: Research in Applied Mathematics. Wiley, Chichester (1994)
54. Komornik, V., Loreti, P.: Fourier Series in Control Theory. Springer Monographs in Mathematics. Springer, New York (2005)
55. Koosis, P.: Introduction to H^p Spaces. Cambridge Tracts in Mathematics 115, 2nd edn. Cambridge University Press, Cambridge (1998)
56. Korobov, V.I., Sklyar, G.M.: Time-optimality and the power moment problem. Math. USSR-Sb. **62**, 185–206 (1989)
57. Krabs, W.: On Moment theory and Controllability of One-Dymensional Vibrating Systems and Heating Processes. Lect. Notes in Control and Information Sciences 173. Springer, Berlin (1992)
58. Kreĭn, M.G., Nudel'man, A.A.: The Markov Moment Problem and Extremal Problems. Translat. of Mathematical Monographs. AMS, Providence (1977)
59. Lasiecka, I.: Unified theory of abstract parabolic boundary problems–A semigroup approach. Appl. Math. Optim. **6**, 287–333 (1980)
60. Lasiecka, I., Triggiani, R.: A cosine operator approach to $L_2(0, T; L_2(\Gamma))$-boundary input hyperbolic equations. Appl. Math. Optim. **7**, 35–93 (1981)
61. Lasiecka, I., Triggiani, R.: Regularity of hyperbolic equations under $L_2(0, T; L_2(\Gamma))$-Dirichlet boundary terms. Appl. Math. Optim. **10**, 275–286 (1983)
62. Lavanya, R., Balachandran, K.: Controllability results of linear parabolic integrodifferential equations. Diff. Integr. Equat. **21**, 801–819 (2008)
63. Leugering, G.: Exact controllability in viscoelasticity of fading memory type. Appl. Anal. **18**, 221–243 (1984)
64. Leugering, G.: Time optimal boundary controllability of a simple linear viscoelastic liquid. Math. Methods Appl. Sci. **9**, 413–430 (1987)
65. Leugering, G.: On boundary controllability of viscoelastic systems. In: Control of Partial Differential Equations. Lecture Notes in Control and Inform. Sci., 114, pp. 190–220. Springer, Berlin (1989)
66. Lions, J-L.: Contrôlabilité exacte, perturbations et stabilization de systémes distribués. Recherches en Mathé matiques Appliqué e 9, vol. 1, Masson, Paris (1988)
67. Liu, Z., Zheng, S.: Semigroups Associated with Dissipative Systems. Chapman & Hall/CRC Research Notes in Mathematics 398. Chapman & Hall/CRC, Boca Raton (1999)
68. Loreti, P., Sforza, D.: Reachability problems for a class of integro-differential equations. J. Diff. Equat. **248**, 1711–1755 (2010)
69. Loreti, P., Pandolfi, L., Sforza, D.: Boundary controllability and observability of a viscoelastic string. SIAM J. Control Optim. **50**, 820–844 (2012)

70. Micu, S., Zuazua, E.: An Introduction to the Controllability of Partial Differential Equations. In Quelques questions de théorie du contrôle Sari, T. ed., Collection Travaux en Cours, Hermann, Paris 69–157 (2004)
71. Mikhailov, V.P.: Partial Differential Equations. Mir, Moscou (1978)
72. Naylor, A.W., Sell, G.R.: Linear Operator Theory in Engineering and Science. Applied Mathematical Sciences 40, 2nd edn. Springer, New York (1982)
73. van Neerven, J.M.A.M.: Null controllability and the algebric Riccati equation in Banach spaces. SIAM J. Control Optim. **43**, 1313–1327 (2005)
74. Pandolfi, L.: The controllability of the Gurtin-Pipkin equation: a cosine operator approach. Appl. Math. Optim. **52**, 143–165 (2005) (a correction in Appl. Math. Optim. **64**, 467–468 (2011))
75. Pandolfi, L.: Riesz systems and controllability of heat equations with memory. Integr. Eqn. Oper. Theory **64**, 429–453 (2009)
76. Pandolfi, L.: On-line input identification and application to Active Noise Cancellation. Annu. Rev. Control **34**, 245–261 (2010)
77. Pandolfi, L.: Riesz systems and moment method in the study of heat equations with memory in one space dimension. Discrete Contin. Dyn. Syst. Ser. B. **14**, 1487–1510 (2010)
78. Pandolfi, L.: Riesz systems and an identification problem for heat equations with memory. Discrete Contin. Dyn. Syst. Ser. S **4**, 745–759 (2011)
79. Pandolfi, L.: Boundary controllability and source reconstruction in a viscoelastic string under external traction. J. Mat. Anal. Appl. **407**, 469–479 (2013)
80. Pandolfi, L.: Traction, deformation and velocity of deformation in a viscoelastic string. Evol. Equ. Control Theory **2**, 471–493 (2013)
81. Pandolfi, L.: Cosine operator and controllability of the wave equation with memory revisited, submitted, arXiv:1407.3706
82. Pandolfi, L.: Sharp control time for viscoelastic bodies. In print, J. Integr. Eqn. Appl., arXiv:1305.1477
83. Pandolfi, L., Priola, E., Zabczyk, J.: Linear Operator Inequality and Null Controllability with Vanishing Energy for boundary control systems. SIAM J. Control Optim. **51**, 629–659 (2013)
84. Pazy, A.: Semigroups of linear operators and applications to partial differential equations. Applied Mathematical Sciences 44. Springer, New York (1983)
85. Prüss, J.: Evolutionary Integral Equations and Applications. Monograph in Mathematics 87. Birkhäuser, Basel (1993)
86. Rolewicz, S.: On universal time for the controllability of time-depending linear control systems. Studia Math. **59**, 133–138 (1976/77)
87. Romanov, I., Shamaev, A.: Exact controllability of the distributed system, governed by string equation with memory. J. Dyn. Control Syst. **19**, 611–623 (2013)
88. Schmidt, E.J.P.G., Weck, N.: On the boundary behavior of solutions to elliptic and parabolic equations–with applications to boundary control of parabolic equations. SIAM J. Control Optim. **16**, 593–598 (1978)
89. Smirnov, V.: Course de Mathématiques Supérieures. Tome II, MIR, Moscou (1970)
90. Soriano, J.A.: Exact controllability of the wave equation. Rev. Mat. Univ. Complut. Madrid **8**, 459–493 (1995)
91. Sova, M.: Cosine operator functions. Rozprawy Mat. **49**, 1–47 (1966)
92. Tikhonov, A.N., Samarskii, A.A.: Equations of Mathematical Physics. Macmillan, New York (1963)
93. Tucsnak, M., Weiss, G.: Observation and Control for Operator Semigroups. Birkhäuser advanced texts, Basel (2009)
94. Wang, J.-M., Guo, B.-Z., Fu, M.-Y.: Dynamic behavior of a heat equation with memory. Math. Methods Appl. Sci. **32**, 1287–1310 (2009)
95. Weck, N.: Über das Prinzip der Eindeutigen Fortsetzbarkeit in der Kontrolltheorie. In: Optimization and optimal control, Lecture Notes in Mathematics 477, pp. 276–284. Springer, Berlin (1975)

96. Yan, J.: Contrôlabilité exacte pour des systèmes à mémoire. Rev. Mat. Univ. Complut. Madrid **5**, 319–327 (1992)
97. Yong J., Zhang, X.: Exact controllability of the heat equation with hyperbolic memory kernel. In: Control of Partial Differential Equations, Lecture Notes in Pure and Applied Mathematics 242, pp. 387–401. Chapman & Hall/CRC, Boca Raton (2005)
98. Young, R.: An Introduction to Nonharmonic Fourier Series, 1st edn. Academic Press Inc, San Diego (2001)

Series Editors' Biographies

Tamer Başar is with the University of Illinois at Urbana-Champaign, where he holds the academic positions of Swanlund Endowed Chair, Center for Advanced Study Professor of Electrical and Computer Engineering, Research Professor at the Coordinated Science Laboratory, and Research Professor at the Information Trust Institute. He received the B.S.E.E. degree from Robert College, Istanbul, and the M.S., M.Phil, and Ph.D. degrees from Yale University. He has published extensively in systems, control, communications, and dynamic games, and has current research interests that address fundamental issues in these areas along with applications such as formation in adversarial environments, network security, resilience in cyber-physical systems, and pricing in networks.

In addition to his editorial involvement with these *Briefs*, Başar is also the Editor-in-Chief of *Automatica*, Editor of two Birkhäuser Series on *Systems & Control* and *Static & Dynamic Game Theory*, the Managing Editor of the *Annals of the International Society of Dynamic Games* (ISDG), and member of editorial and advisory boards of several international journals in control, wireless networks, and applied mathematics. He has received several awards and recognitions over the years, among which are the Medal of Science of Turkey (1993); Bode Lecture Prize (2004) of IEEE CSS; Quazza Medal (2005) of IFAC; Bellman Control Heritage Award (2006) of AACC; and Isaacs Award (2010) of ISDG. He is a member of the US National Academy of Engineering, Fellow of IEEE and IFAC, Council Member of IFAC (2011–2014), a past president of CSS, the founding president of ISDG, and president of AACC (2010–2011).

Antonio Bicchi is Professor of Automatic Control and Robotics at the University of Pisa. He graduated at the University of Bologna in 1988 and was a postdoc scholar at M.I.T. A.I. Lab between 1988 and 1990. His main research interests are in:

- dynamics, kinematics and control of complex mechanichal systems, including robots, autonomous vehicles, and automotive systems;
- haptics and dextrous manipulation; and
- theory and control of nonlinear systems, in particular hybrid (logic/dynamic, symbol/signal) systems.

© The Author(s) 2014
L. Pandolfi, *Distributed Systems with Persistent Memory*, SpringerBriefs in Control,
Automation and Robotics, DOI 10.1007/978-3-319-12247-2

He has published more than 300 papers on international journals, books, and refereed conferences. Professor Bicchi currently serves as the Director of the Interdepartmental Research Center "E. Piaggio" of the University of Pisa, and President of the Italian Association or Researchers in Automatic Control. He has served as Editor in Chief of the Conference Editorial Board for the IEEE Robotics and Automation Society (RAS), and as Vice President of IEEE RAS, Distinguished Lecturer, and Editor for several scientific journals including the *International Journal of Robotics Research*, the *IEEE Transactions on Robotics and Automation*, and *IEEE RAS Magazine*. He has organized and co-chaired the first WorldHaptics Conference (2005), and Hybrid Systems: Computation and Control (2007). He is the recipient of several best paper awards at various conferences, and of an Advanced Grant from the European Research Council. Antonio Bicchi has been an IEEE Fellow since 2005.

Miroslav Krstic holds the Daniel L. Alspach chair and is the founding director of the Cymer Center for Control Systems and Dynamics at University of California, San Diego. He is a recipient of the PECASE, NSF Career, and ONR Young Investigator Awards, as well as the Axelby and Schuck Paper Prizes. Professor Krstic was the first recipient of the UCSD Research Award in the area of engineering and has held the Russell Severance Springer Distinguished Visiting Professorship at UC Berkeley and the Harold W. Sorenson Distinguished Professorship at UCSD. He is a Fellow of IEEE and IFAC. Professor Krstic serves as Senior Editor for *Automatica* and *IEEE Transactions on Automatic Control* and as Editor for the Springer series *Communications and Control Engineering*. He has served as Vice President for Technical Activities of the IEEE Control Systems Society. Krstic has co-authored eight books on adaptive, nonlinear, and stochastic control, extremum seeking, control of PDE systems including turbulent flows, and control of delay systems.

Index

© The Author(s) 2014
L. Pandolfi, *Distributed Systems with Persistent Memory*, SpringerBriefs in Control,
Automation and Robotics, DOI 10.1007/978-3-319-12247-2